into the **Hidden environment**

into the HIDDEN ENVIRONMENT

The Oceans

Keith Critchlow

Special illustrations painted by David Nockels

A Studio Book · The Viking Press · New York

Published in 1973 by The Viking Press, Inc.
625 Madison Avenue New York NY 10022

Published simultaneously in Canada by
The Macmillan Company of Canada Limited

Made by Roxby Press Productions
55 Conduit Street London W1R 9FD
Editor Michael Wright
Picture research Elizabeth Lake
Design and art direction Ivan and Robin Dodd

Printed in Great Britain by Oxley Press Limited

ISBN: 670-40029-7

Library of Congress Catalog Card No: 72-81672

Contents

Introduction

VAST, mysterious gorges deep enough to swallow Everest with 6,000 feet to spare; cliff escarpments ten times higher than any seen by man; ridges and mountain ranges tens of thousands of miles long, virtually unattacked by erosion, yet immaculately blanketed in the ever-falling 'snow' of the sediments; deserts of a flatness that make the Sahara seem like a sculptured garden: these are among the undreamed-of features of the hidden undersea environment uncovered by modern deep-sea science. And yet this science is still in its infancy. The fact of the matter is that men have never seen the majority of the planet's surface. This book aims to project into the submarine world, to whet the appetites of future explorers, and to present the issues involved in our oceanic destiny.

The motives are clear: the oceans are vital to the very existence of life. If men are to survive long enough actually to see these magnificent undersea landscapes, we need to understand the oceans and to appreciate the urgency of preserving the delicate oceanic balance, since it is the very balance of life itself. Every tiny bead of perspiration brought out by the Sun on any one of us has come from and will return to the ocean. Its saltiness, and that of the bloodstream from whence it came, both echo the composition of the seas when, millions of years ago, our predecessors emerged from their waters. The oceans were the very womb and cradle of life; the seventy-one per cent water of the human body is a constant reminder of these origins.

The qualities and characteristics of water are in themselves nothing short of miraculous. The fact that ice floats, that sap can almost deny gravity in a tall tree, that almost all chemicals can be dissolved in water, and that it is the most impressionable of substances, all are directly responsible for an aspect of the maintenance of life. And the Earth, because of its unique harmonic distance from the Sun, is able to maintain the correct balance of temperatures for water to co-exist in its three characteristic states: solid ice, liquid and vapour.

It is because of the complete dependence of life on water that people are linked so inextricably to the great cyclic water movements between and within the atmosphere, the land and the gigantic world ocean. The seas have been paralleled with the circulatory systems of a person's body, distributing nourishment, regulating temperatures, and carrying away various forms of waste matter. There is a parallel, too, between the great water cycles of the planet and the fact that living organisms rely not merely on *containing* water, but on having water pass constantly through their bodies.

It may seem strange to talk of the Earth-organism, but probably the most relevant way to think of the planet as a whole and the oceans in particular is as a living, organic system. Such an organic model helps to emphasize the complete interdependence of all the products of the planet over the vast millennia of its history. Most important of all, it is essential to remember that man is a part and product of the Earth. Although it is our home, the Earth and its ocean cannot *belong* to us; we are inextricably part of it and it is part of us. The traditional phrase 'mother Earth' unconsciously emphasizes this interrelationship.

If the Earth is some kind of organism, then man is its most self-conscious and active constituent. He is the thinking element; he expresses the purposive part, and is responsive in the greatest number of ways. He is, in a sense, the nervous system of the planet, and it has become clear that the rest of the Earth-organism relies utterly for its continued functioning and survival on man's co-operative behaviour.

That is why the subject-content of this book is so important. If mankind has become capable of making or breaking this delicate organic balance, then the more that can be understood about the structure, flow, balance and details of the physical and biological forces making up the collective planetary 'body' the better. Since man can be regarded as the thinking element of the whole system, his special responsibility is to base his actions on understanding. It is a challenge as to whether he can understand fully and act responsibly before myopic and selfish exploitation of the Earth-organism spoils the system for all.

And yet the attitude needed is only that of a sound peasant farmer to his land: an attitude of respect for the timing of natural processes and an understanding of the preciousness of the soils, of the climatic subtleties and the interdependency of the life cycles. With a similar attitude to the oceans, we might harvest much of their potential with a new sensitivity – a sensitivity born of enlightened self-interest but heavy with the knowledge of what happens when we indiscriminately upset the chemical balance in the natural food chains.

As an example of the understanding that is necessary, take the role of the oceans as the planet's cesspool. For as long as life has existed, much of the excrement and dead remains of living things have found their way directly or indirectly back to the sea, whether through man-made sewage systems or soil erosion. Once rid of, the wastes are forgotten, but the oceans are not mere containers for them. Just as bacteria in a cesspool digest sewage, so are the

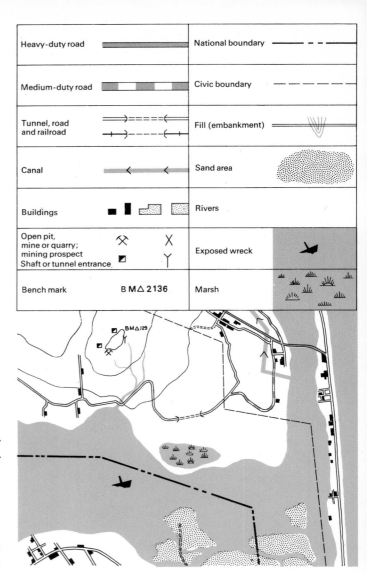

Heavy-duty road		National boundary	
Medium-duty road		Civic boundary	
Tunnel, road and railroad		Fill (embankment)	
Canal		Sand area	
Buildings		Rivers	
Open pit, mine or quarry; mining prospect Shaft or tunnel entrance		Exposed wreck	
Bench mark	B M△ 2136	Marsh	

wastes in the seas broken down, transformed and redistributed as food supplies for the aquatic billions.

Traditionally, man has worked with the principle that all organically based materials, when they become waste, will automatically become broken down – so out of sight, out of mind. People's greatest environmental mistake is perhaps to think that this 'out of sight out of mind' attitude is equally valid for modern industrial wastes. If the natural digestive system of the planet cannot deal with them – as in the cases of DDT, mercury and lead – then they will find their way back to poison us and a great deal of the system on the way. Today we can see that we are ourselves only one link in a living chain and part of a series of inevitable cycles. It is a contradiction in terms to speak of 'throwing away' something. It must go somewhere; if it cannot be broken down naturally it will be returned to us in some form or other.

Can any of us say we know how industrial wastes are assimilated by the planetary organism? If we cannot foresee how the oceans are going to digest our effluents and industrial garbage, then we will have to insist – even through legislation – that inventors spend as much ingenuity on designing 'digesting machines' for neutralizing by-products, or on ways of re-using a discarded product, as on the original invention itself. An invention in future will have to pass an 'ecology test' to see that it has an ultimate benefit over the whole cycle, including release of the materials for re-use. Eventually, the industrial yardstick of cost-benefit will have to take in every stage in this fundamental cycle, or there will be no benefit at whatever cost. The oceans and their many interpenetrating functions exhibit these principles of use, disposal and re-use in a masterly way.

The task of searching, scanning and exploring the sea beds is immense and exciting – but dark and alive with hazards. The aim of this book is to take you into new areas – part known, part still hidden. Where known facts are scarce, it will be essential to indulge in a certain amount of informed imagination. But it can only be a matter of time before we can see for ourselves, as scanning technology catches up with mankind's need to know.

Only when a man has really been challenged to 'see for himself' has he ever gained true clarity and sound knowledge. The need to know is producing vast improvements in oceanographic techniques. The darkness of the great depths can, for example, be overcome by using sound waves instead of light. More and more sophisticated sonar and other devices will eventually be able to recreate the

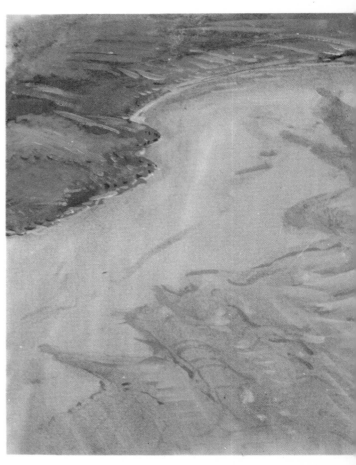

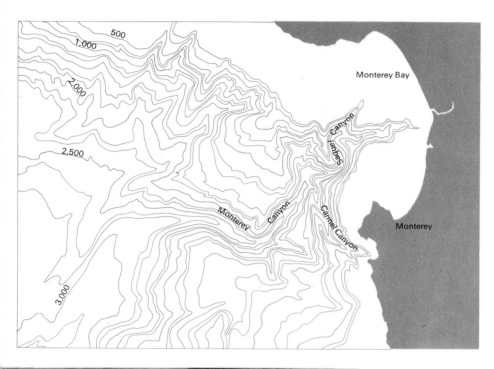

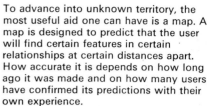

To advance into unknown territory, the most useful aid one can have is a map. A map is designed to predict that the user will find certain features in certain relationships at certain distances apart. How accurate it is depends on how long ago it was made and on how many users have confirmed its predictions with their own experience.

far left The most familiar, detailed and generally reliable maps are those of the land, where various conventional signs are used to indicate topographical features – natural or man-made. Maps – or charts – of the seas are vital for safe navigation, but details of the sea bed are critical only in the shallower waters, and charts tend to be most detailed and accurate in these areas. **left** As oceanographers gather more information about the shape of the sea bed, detailed contour maps can be produced. This shows Monterey Bay, on the coast of California. **below left** This perspective drawing of the same area shows the dramatic underwater canyons revealed by the contour map (see also pages 56 and 57). Such detailed knowledge of the ocean bed exists for only a few areas of the globe, and, in any case, oceanic features change. Just as a new road may be built, so can an underwater volcano appear or disappear. Turbulence storms which form the canyons and the ever-falling snow of the sediments alter the shape of the sea bed. At a slower rate of change, new ocean beds are being formed and old destroyed by the movements of continents. And as our understanding of the forces moulding the planet develops, so the significance of various oceanic features becomes clearer. Thus not only are the features themselves changing, but so is man's view of them. It is in this context of change that the facts about the oceans presented in this book must be viewed.

breathtaking features of the sea bed with startling clarity on a screen. Only then will it be possible to see the full pattern of unity behind the diverse systems and features.

For the sake of familiarity, the oceans can be paralleled with the air-ocean that mankind is so used to inhabiting. Our natural habitat is the bottom of an ocean of air, and in exploring the sea-ocean we are like space-travellers landing in from outer space. In dropping in on this great theatre of life and scene of the most dramatic topological features of the whole planet, the traveller to 'inner space' has a wealth of breathtaking unexplored features to draw him on.

Turbulence and local concentrations within the clouds of dust and gas filling space could easily have drawn the matter into coagulating masses under the force of gravity. In this way the Sun could have been formed and through the force of its gravitational compression ignited its own thermonuclear furnace. But, as a result of magnetic forces, a disk of cold gas and dust would have been left spinning around it.

Coagulations within the disk could have drawn the heavier matter together to form the planets while most of the lighter chemical elements remained dispersed. If our planet was born in this way, similar planetary systems must be quite common in the universe.

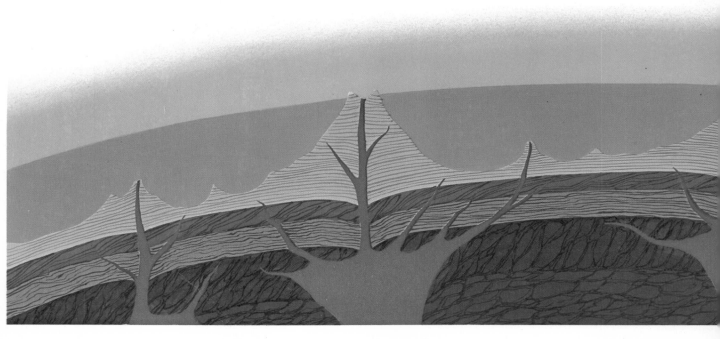

IN comparison with the variable fortunes of human existence, there seems nothing more permanent than our planetary home. And in spite of day-to-day disturbances on their surface – currents, tides, storms – the oceans seem an immutable feature of our surroundings. Yet they do have a history of change, as does the whole planet, and their apparent permanence is more a reflection of the brevity of human life than of the physical world's resistance to alteration. The nature of the oceans today tells us a great deal about upheavals in their past history and points to a continuing story of change in the future.

All physical phenomena are apparently subject to the same rules of existence as human life itself, if on a far greater time-scale. All pass through the stages of conception, birth, growth, maturity, decay, death and reabsorption. To the best of our knowledge, this is true not only of the oceans but of the planet, of the whole Solar System, of the Milky Way of which our Sun forms but a minute part, of all the clusters of galaxies in the universe, and indeed of the universe itself. So in dealing with the birth of the oceans it is inevitable that the issue of greater origins arises.

Possibly there is no more profound way of expounding this idea of the birth of the natural universe than in the poetic words of the Ancient Chinese philosopher Lao Tzu, who wrote 2,500 years ago:

There was something undifferentiated and yet complete, which existed before Heaven and Earth. Soundless and form-less, it depends on nothing and does not change. It operates everywhere and is free from danger. It may be considered the mother of the universe. I do not know its name; I call it Tao (Way). If forced to give it a name, I shall call it Great. Now being great means functioning everywhere. Functioning everywhere means far-reaching. Being far-reaching means returning to the original point.

These words show a remarkably intuitive understanding of concepts that occur in modern cosmology and relativity – such concepts as the conservation of mass-energy; the view of the universe as an expanding, contracting or steady-state system; and the double curvature of space-time. Above all, Lao Tzu emphasizes the dangers of making final definitions and statements of fact. However, the function of science is to put in order the facts of experience, so an attempt must be made, based on the most up-to-date knowledge.

Creation within the universe itself seems to progress in a definite sequence, beginning with the unorganized chaos of formlessness. From this dispersed state, gravitational instabilities arise so that the dust and unorganized matter become less uniformly spread out, and begin to coagulate towards centres of evolving organization or order. Stars apparently form out of these rotating gaseous globules, and evolve by creating new substances from the basic hydrogen

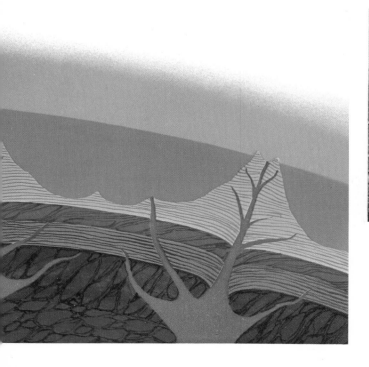

above and left Volcanoes erupting, pouring molten mantle out through the Earth's solid crust: From such alchemical retorts as these, water vapour and gases were distilled out of the primitive planet's interior to form the atmosphere and oceans. The total water content of the Earth has apparently been finite and constant since its creation.

under unimaginably intense heat caused by the massive compressive forces at the centre. In some cases, a star in the late stages of its evolution becomes unstable and explodes with the colossal force of a supernova. In the nuclear inferno of such exploding stars are created the heavier chemical elements, to be flung far out into space.

Two ways of explaining the birth of the Solar System must be mentioned. The first, put forward some 200 years ago by French mathematician Pierre Laplace, envisages the Sun and planets evolving from a rotating cloud of cosmic gas. During compression under the influence of its own gravitational attraction, this is seen as forming the Sun at the centre and dropping rings of matter on the way which eventually coagulated into the separate planets. There are various objections to Laplace's explanation, among them the fact that the orbits of the planets would be expected to lie in the same plane as the spin of the Sun, just as the drop of mud flung off a spinning bicycle wheel keeps travelling in the same plane. The planets do not obey this prediction.

In the late 1920s, Sir James Jeans countered Laplace's theory with another based on the principle of tides. He suggested that our planetary system had two parents, the other being a passing star. This momentous event drew out of our Sun a cigar-shaped filament of gas, which co-agulated into the multiple system existing today. It has been objected that neither of these two theories can survive the fact that the chemical composition of the planets is not the same as that of the Sun. One solution has been proposed by Fred Hoyle and R. A. Lyttleton. They see the Solar System originating as a pair of stars, one of which exploded to create the debris which eventually became a system of planets circling the one remaining – the Sun.

Just as there are differing opinions of how a nebulous cloud of cosmic gas progressed into a harmoniously balanced system of planets, so there are differing opinions as to how the oceans were born. On one aspect, there is general agreement: that the water content of the planet has remained virtually constant from the earliest days. However, whether this water condensed slowly into rain from very hot vaporous clouds, or whether it was distilled from the Earth's interior by volcanoes, is not yet clear. The volcanic theory suggests that the heavy metals were attracted to the centre of the molten core, while the lighter components, including water, were squeezed out of the crust during the cooling process through the volcanic valves. This latter is the more dynamic picture, and geology in recent years has undergone a revolution in thought, with the emergence of the dynamic view of global feature-building. In particular, the creation of the basins into which the planets' water drained is now seen as an evolutionary, continuing process.

The drifting continents

It is much more than a truism to say that the oceans fill the spaces between the continents. For the picture that has emerged since 1960 from the feverish activities of geophysicists is one of floating, drifting and (in some cases) colliding continents, of spreading sea beds and growing oceans, of the Earth's crust being swallowed into the mantle in one place to be reborn elsewhere, of the 'solid' rock being churned by currents – albeit on a much greater time-scale than we can easily comprehend.

The idea that the continents have been moved around on the surface of the Earth, like so many icebergs on a polar sea, is not new. But the first comprehensive theory of continental drift was put forward in 1912 by a German meteorologist, Alfred Wegener. Greeted at first with disdain, even ridicule, Wegener's theory has re-emerged –

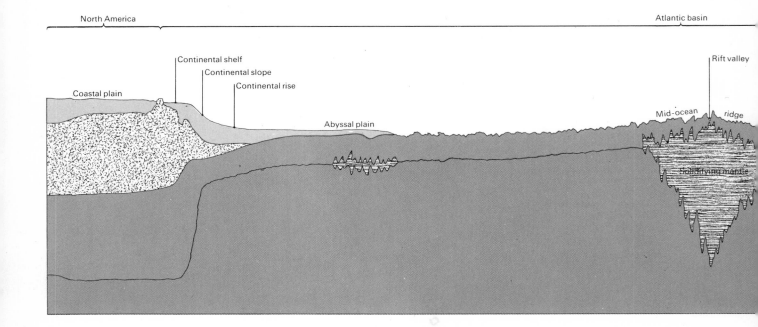

North America   Atlantic basin

Continental shelf
Continental slope
Continental rise

Coastal plain

Rift valley

Abyssal plain

Mid-ocean   ridge

Solidifying mantle

though considerably modified – as the cornerstone of the new geology. The latest developments indicate that the surface of the Earth is built of six major plates – five of them carrying continental land masses – and a number of other minor ones. The largest plate of all is the Pacific, which has no continent floating on it. The others are the Antarctic plate (carrying Antarctica), the African plate (Africa), the Indian plate (Arabia, India and Australia), the Eurasian plate (Europe and the bulk of Asia), and the American plate (North and South America). Between these, ocean floor is constantly being born at the centres known as the mid-ocean ridges, and is being buried for recirculation in the deep trenches.

Wegener's theory was probably disregarded for so many years because of a deep-rooted subconscious fear of relinquishing one of the great axioms of stability and permanence – the solidity of mother Earth. However, the new generation of geologists have been swung in favour of continental drift by two linked discoveries: the seismic liveliness of the mid-ocean ridges, and the fossilized 'tape recordings' of changes in the Earth's magnetic field impressed in the ocean bed on either side of these ridges. At various times in the Earth's history – at intervals of approximately 12 million years – the direction of the planet's magnetic field has dramatically reversed. North magnetic pole has become south, and south pole north. Any new crust that is crystallized from the mantle carries the fossilized magnetic imprint of the field at that instant. And either side of the mid-ocean ridges lie stripes of oppositely magnetized rock, parallel to the ridge, with increasingly old rock the farther one moves from the ridge – indicating a progressive spreading outwards. More recently, this expansion has been actually measured in Iceland, where the Mid-Atlantic Ridge rises above sea level. And the theory has been extended to encompass mountain-building, now seen as a result of lateral pushings and continental collisions.

The illustrations of continental drift on page 15 are based on the 1922 version of Wegener's proposals. Recent developments have however indicated that India drifted north to meet the Eurasian plate independently, the impact of these continents resulting in the drama of the Himalayas. One result of this impact was to raise sedimentary rocks – originally formed from sea-bed silts – as high as 19,000 feet. It was a group of scientists under Professor P. M. S. Blackett of London University who first trailed the passage of India after its parting with Antarctica up to its final wedging place. Using fossil magnetism, they estimated a travelling speed of about three centimetres per year. More recently, American geologist Bruce Heezen found evidence that the Maldive Ridge in the Indian Ocean was left behind by the passage of India in the early Eocene epoch, some 50 million years ago. It is also possible that the East Indian Ridge (or Ninety-East Ridge) could be regarded as a 2,500-mile-long skid mark left by the eastern edge of the continent.

The power to move continents

As a descriptive theory, the idea of continental drift has been a resounding success, but it begs a massive question. What force can shift the continents as they float on the heavier mantle? Scientists now believe that world-wide churning of the crust's under-surface by convection currents in the mantle could be the answer. Convection currents occur where hot, light material – gas or liquid – rises, while cooler, heavier material sinks to replace it. If the main mountain ranges of the planet demarcate an almost continuous pressure belt, and the mid-ocean ridges

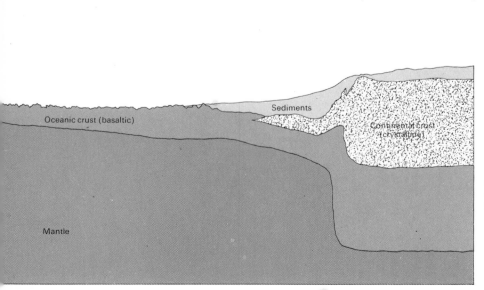

The ocean bed grows: A revolution in geology has resulted from confirmation that many oceans are steadily widening as the sea bed spreads on either side of the mid-ocean ridge. The most widely-held theory sees this as caused by convection currents in the plastic mantle. As a result, new crust is formed and pushed up in the rift valley of the mid-ocean ridge, while elsewhere crust is swallowed again in the deep ocean trenches. But the facts could also be explained if the Earth were slowly expanding.

14

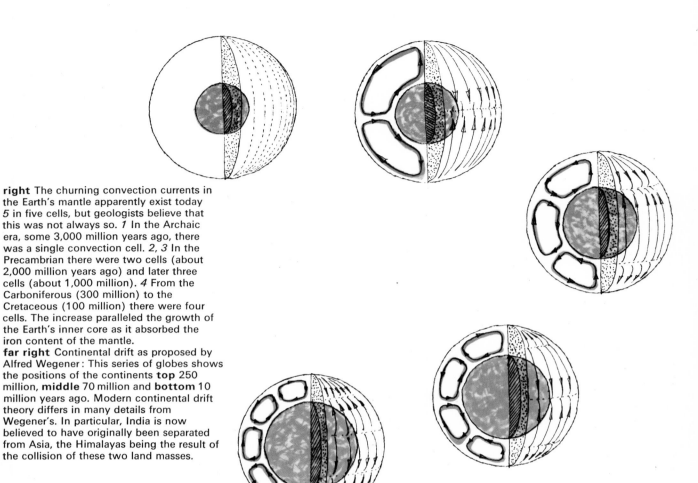

right The churning convection currents in the Earth's mantle apparently exist today 5 in five cells, but geologists believe that this was not always so. 1 In the Archaic era, some 3,000 million years ago, there was a single convection cell. 2, 3 In the Precambrian there were two cells (about 2,000 million years ago) and later three cells (about 1,000 million). 4 From the Carboniferous (300 million) to the Cretaceous (100 million) there were four cells. The increase paralleled the growth of the Earth's inner core as it absorbed the iron content of the mantle.
far right Continental drift as proposed by Alfred Wegener: This series of globes shows the positions of the continents top 250 million, middle 70 million and bottom 10 million years ago. Modern continental drift theory differs in many details from Wegener's. In particular, India is now believed to have originally been separated from Asia, the Himalayas being the result of the collision of these two land masses.

a complementary tension seam, then the connections between these phenomena indicate the existence of world-encircling convection patterns. Also, the mid-ocean ridges and the trenches where the sea floor is swallowed up are alive with volcanoes, where fluid mantle is disgorged onto the surface.

As the illustrations on this page show, there are apparently five churning convection 'cells' beneath the planet's crust. This was discovered by the Dutch scientist F. A. Vening Meinesz, who originally put forward the convection theory after finding unexpected changes in the Earth's gravity during a submarine survey of the Pacific Ocean in 1951. The moving surface plates are seen as the horizontal elements of the convection cells; the bedrock beneath our feet is in constant circulation.

It should be mentioned, however, that the convection-current idea is not the only suggestion for the force behind the drifting continents. Bruce Heezen among others has proposed an even more radical idea. If the Earth is imagined contracted to half its present diameter, the result is to all but assemble the continents into a complete skin. Has the Earth then, over a period of 3,000 to 4,000 million years, gradually grown to its present size, tearing apart its crust to leave gaping ocean basins in the process? Cosmologist Fred Hoyle has suggested that such an expansion might be the result of a slow but inexorable weakening of the force of gravity over the aeons of time. One attractive aspect of this

theory is the difficulty of disproving it. The rate of expansion of the planet would amount to one millimetre per year – a precise amount, but so far impossible to measure.

Where the crust-plates meet
Whether one chooses to visualize the continents spreading across an expanding globe or circulating on a churning mantle, the general theory of continental drift depends very largely on the concept of moving plates. The study of the plates' behaviour is known as plate tectonics. All the six major plates are apparently created in the same way – through material from the mantle rising up along the mid-ocean ridges – but their termination varies. Sometimes there is a massive compressional buckling and folding, such as that which formed the Himalayas or the Alps. Sometimes one plate dives down below its neighbour, forming a deep ocean trench, such as those in mid-Pacific. Sometimes the result is both a trench and a mountain range, as at the western edge of South America. A further possibility found at the junction of some plates is the sliding of one plate past another in a massive transform fault. All these patterns were apparently created due to forces of heat and convection within the mantle, but the exact mechanisms and the reasons for the differences remain shrouded in obscurity at present.

Nor is there much detailed knowledge of what exactly is happening at the plate edges, except where the Pacific and

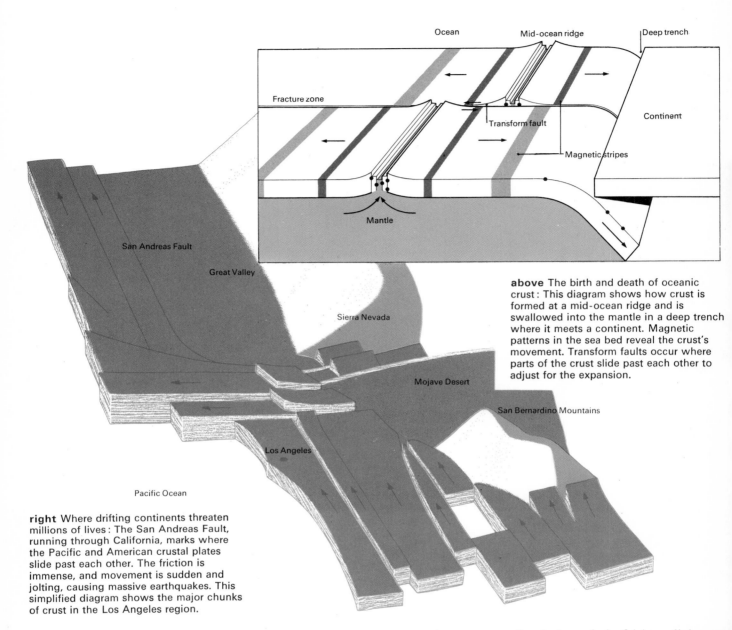

Ocean — Mid-ocean ridge — Deep trench

Fracture zone

Transform fault

Continent

Magnetic stripes

Mantle

San Andreas Fault

Great Valley

Sierra Nevada

Mojave Desert

San Bernardino Mountains

Los Angeles

Pacific Ocean

above The birth and death of oceanic crust: This diagram shows how crust is formed at a mid-ocean ridge and is swallowed into the mantle in a deep trench where it meets a continent. Magnetic patterns in the sea bed reveal the crust's movement. Transform faults occur where parts of the crust slide past each other to adjust for the expansion.

right Where drifting continents threaten millions of lives: The San Andreas Fault, running through California, marks where the Pacific and American crustal plates slide past each other. The friction is immense, and movement is sudden and jolting, causing massive earthquakes. This simplified diagram shows the major chunks of crust in the Los Angeles region.

American plates meet along the Californian coast. Here – partly prompted by the need to understand and, if possible, tame the potentially massive destructive force of the San Andreas fault – detailed studies were undertaken in the late 1960s and early 1970s. The findings are complex, but the detective work in making them is exciting. Some of the discoveries of the team under Professor Don L. Anderson are shown diagrammatically on this page.

The interacting global systems
The dynamic concepts of the new geology deserve to be extended from a mere consideration of continents in motion (with the implied changes in the oceans) to a new view of the whole planetary system and its components. These are commonly seen as a series of 'spheres': the solid *lithosphere*, the liquid *hydrosphere*, the *atmosphere*, and the living *biosphere*. But these are now known all to be in a constant state of flux – albeit on widely differing time-scales – so why not emphasize the fact with a series of new terms? Because of the vital link between all life and a supply of

water, the concept of a *hydrocycle* is fairly well known, emphasizing the constant movement of solid, liquid and vaporous water between the land, sea and air. It is a very small extension of this terminology to use the word *biocycle* to emphasize the interdependent life cycles of animals and plants. Extending this to the other major planetary spheres, we have the terms *lithocycle* for the circulating bedrock and *atmocycle* for the constantly moving airstreams. Finally, the term *thermocycle* might be introduced to relate us to the cosmically rare and vital phenomenon of solar warmth, for it is due only to the Earth's distance from the Sun that water on our planet can coexist in all its forms and thus support life.

These new terms have the value of renewing attitudes which at present compartmentalize the states of matter in their separate spheres. They bring a new viewpoint that stresses the constant interchange and interdependency. Not only is the lithocycle circulating on the grand scale indicated by continental drift, but it is also circulating within the hydrocycle as silts, sediments and vital ingredients for marine life. The concept of an atmocycle stresses not only the air currents but also the exchange of vital gases with the waters and life cycles.

The four states of matter and energy – solid, liquid, gas, radiation – are all very real facts of our experience, and they behave in different ways in the planetary system. But it is necessary to visualize the amazing balance that has evolved between these interpenetrating cycles, both to understand the emergence of the even more amazing phenomenon of life, and – a topic examined in a later chapter – to understand the threat of mankind's wanton, if largely unconscious, interference with this balance through pollution.

The cycles present a complex and rather awe-inspiring picture: the cohesive gravity of a sphere of rock bearing a sphere of water, both subject to tidal forces of Sun and Moon, and a sphere of air virtually transparent to the constant flow of sunlight; the heating of land and sea giving rise to varying distributions of heat in the atmosphere; the global transport of wind and water determined by pressure differences created by these variations; the massive raising of water as vapour, to be deposited (usually within 12 days) on land and sea as rain, sleet, snow or hail; the slow destruction of the lithospheric continents through the action of winds and rains, and the carrying of the particles back to sea; the huge circulations of magnetic fields which thread through all of the spheres and, with the ozone of the upper atmosphere, reduce and filter the wavelengths of sunlight harmful to life; the properties of water that make it so good a storage receptacle for the 40 per cent of solar radiation that arrives as radiant heat; the 'greenhouse effect' of the air in retaining warmth that reaches the planet's surface. All these remarkable and dazzling forces need to be visualized simultaneously, to set the scene for the throwing of the great cosmic switch, the creation of life.

The arrival of life

It is very difficult to tell what exactly the Earth was like when life came into existence, or to tell precisely when it did so. It is also difficult to know what the planetary equilibrium was before life made such enormous changes to the face of the Earth. Recall for a moment that all the known sedimentary rocks on all the continents were built when the seas were already teeming with life; and that the Earth's oldest known rocks – found in Greenland in the late 1960s by New Zealander Vic McGregor – are, at an age of about 3,700 million years, probably no older than the first living things. The utter barrenness of a non-living planet would be almost unimaginable had not man visited the Moon and realized how fantastically hostile in every sense it can be.

Modern biology has gained increasing insights into the structure and conditions of life and its chemical basis, but this, as the late Professor J. D. Bernal emphasized, in no

Studies of sea-bed magnetization supplied the proof of continental drift theory. Periodically, the Earth's magnetic field has completely reversed its direction. New crust formed from the mantle has frozen into it an echo of the Earth's field at the moment of solidification. Thus the direction of magnetization reveals the age of a particular piece of crust. Detailed magnetic surveys give maps of the sea floor like that reproduced here, which shows the eastern Pacific off the California coast. Such maps reveal the details of past crustal movements.

way diminishes the miraculousness of its occurrence. The basic chemical constituents of living things have a beautiful simplicity. In animals, there are six principal categories: water, the universal solvent; carbohydrates, the main source of energy; fats, the main form of stored energy; adenosine phosphates, the means of energy transfer; proteins (composed of some 20 amino-acid building blocks), both the structural material and the controllers of biochemical reactions; and nucleic acids (built from five nucleotide components), the means of storing, reproducing and transferring genetic information. And these six groups of life-chemicals are themselves built up principally from six chemical elements: hydrogen, oxygen, carbon, nitrogen, phosphorus and sulphur. How simple the scheme seems, yet how little it tells us about the essential nature of the spectrum of living species from a humming-bird to a whale, or from a centipede to a graceful giraffe! Yet all these species are but combinations of the basic materials listed above with a few minor, if important, additions.

Spectroscopic analysis of light from the Sun and planets, together with chemical analysis of meteorites, can reveal the relative abundance of the chemical elements in the Solar System. The conclusion is that the six most important elements for life are among the most abundant throughout the whole system. Water, a combination of two of the three commonest elements, is by far the most important basic component. Even most of the trace elements of living matter are found among the Solar System's top twenty elements. Life's non-living chemical heritage is clear.

But how did the life-chemicals and eventually living organisms themselves come into existence? Discounting the suggestion that the planet was 'seeded' with life via meteorites (which begs the question of how *this* life originated), we are left with the conclusion that the first magic combinations of self-reproducing molecules were created out of the 'primordial soup' of the Earth's oceans and atmosphere back in the misty recesses of our planetary history. Views differ on the composition of the early Earth's atmosphere, but one view suggests a combination of hydrogen, helium, methane and ammonia, together with the ubiquitous water. Stanley L. Miller, a graduate student at Chicago University, sent a wave of excitement and speculation through the biological world in 1952 when, by reproducing with electric sparks the action of lightening on such a mixture, he managed to synthesize amino acids in the laboratory.

Since then, many similar experiments have been conducted, some using fierce ultra-violet light (to reproduce unfiltered sunlight) in place of sparks, and many other life-chemicals have been made, including the components of nucleic acids. Amino acids have been formed into protein-like polymers, and – perhaps most significant of all – these polymers have been found to form spherical globules that may represent the forerunners of living cells.

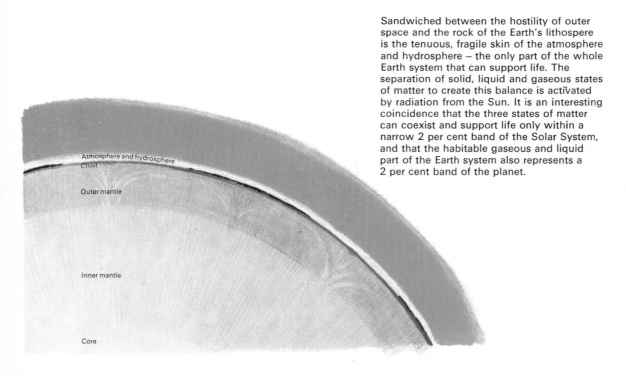

Sandwiched between the hostility of outer space and the rock of the Earth's lithospere is the tenuous, fragile skin of the atmosphere and hydrosphere – the only part of the whole Earth system that can support life. The separation of solid, liquid and gaseous states of matter to create this balance is activated by radiation from the Sun. It is an interesting coincidence that the three states of matter can coexist and support life only within a narrow 2 per cent band of the Solar System, and that the habitable gaseous and liquid part of the Earth system also represents a 2 per cent band of the planet.

Atmosphere and hydrosphere

Crust

Outer mantle

Inner mantle

Core

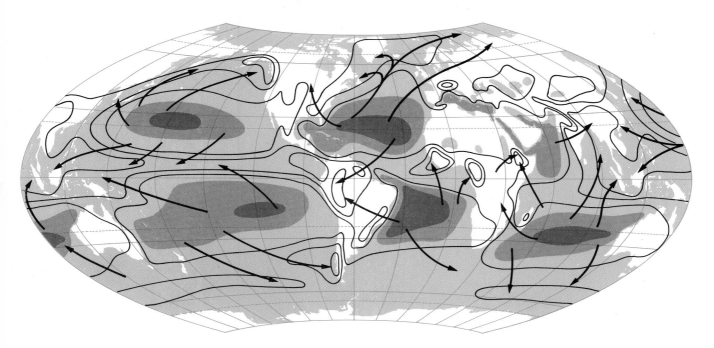

above A massive tonnage of life-sustaining water is constantly distributed from the seas to the land areas of the planet, as shown on this map by the arrows. The darkest tinted areas represent the greatest evaporation. This distribution is the result of intimate cooperation between the liquid and gaseous parts of the planet and the filtered radiation from the Sun.

right The currents of the oceans are far more complex than anyone has dreamed, forming a subtle three-dimensional system that Matthew Maury termed the 'arteries of the planet'. Off the eastern coast of North America, for example, the warm Gulf Stream on the surface is countered by a deeper cold current.

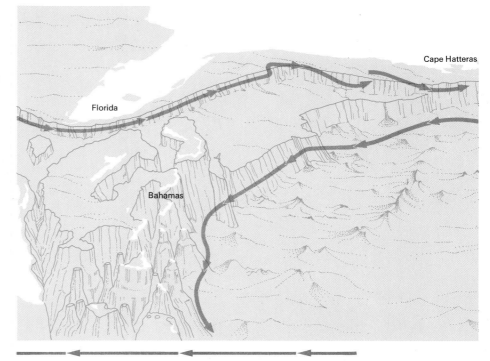

Cape Hatteras

Florida

Bahamas

right In the shallow waters of the continental shelves, offshore winds cause vertical circulation of the waters and upwelling of vital nutrients from the rich sea-bed silts. Such upwelling is directly responsible for the great food production of the continental shelves, for it greatly increases the amount of photosynthesizing phytoplankton, which are the beginning of every marine food chain.

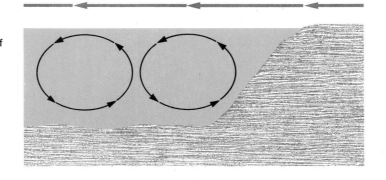

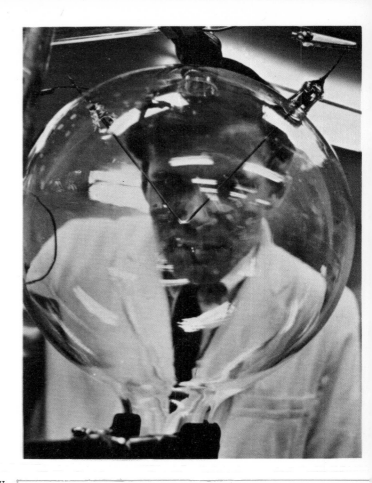

There are still many mysterious, unexplained and certainly as yet unreproduced steps between these primitive chemical systems and the emergence of the first living things. In particular, the greatest advance of all needs to be explained: how proteins and nucleic acids formed the productive partnership that is the cornerstone of life. But some scientists are beginning to look upon life as an inevitable consequence of the conditions existing on Earth some 4,500 million years ago, rather than as a happy accident.

There is little fossil evidence with which to trace the early stages of evolution, and none older than about 2,000 million years. Undoubtedly, the most significant advance as far as man is concerned was the emergence of abundant animal life about 600 million years ago. Evidence suggests that this marks the point when the earlier plant organisms had produced enough oxygen to transform the Earth's atmosphere and support animal respiration. Another result of oxygen-production was to generate a protective blanket of ozone (oxygen atoms linked in threes) in the upper atmosphere, thus cutting down the harsh ultra-violet light of the Sun. The earliest algae and bacteria, like their present-day descendants, must have been resistant to this radiation, but even they could probably survive only in deep water or under rocks. The protective ozone allowed the colonization of the shallower waters and dry land, but for a long, long time the oceans were the only inhabited regions.

The majestic route from the first bacteria to upright, fully self-conscious man is far too grand to be neatly tied up in a watertight theory. However, we have to use the best mechanisms for our understanding to date – and developments of Darwin's theory of evolution remain the most consistent we have. Diversification is seen as life's solution to the problem of mastering the hostilities of the environ-

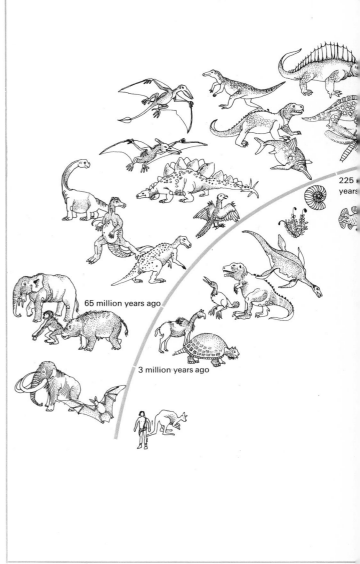

225
years

65 million years ago

3 million years ago

left Did life begin in a lightning-flash?
A spark passes through a mixture of gases
such as may have existed in the primitive
Earth's atmosphere, and amino acids and
other chemical substances fundamental to
living systems are created. Man reproduces
the first steps to life in the laboratory.
below Or was the Earth 'seeded' with life
from outer space? This photomicrograph of
a meteorite shows objects looking remarkably
like fossil algae. bottom The procession
of life-forms on Earth: The period in which
advanced life has existed represents a very
small part of the planet's total history.
Man's emergence is a mere fleeting moment.

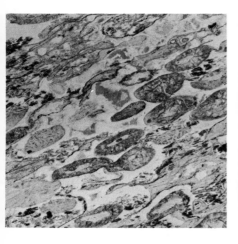

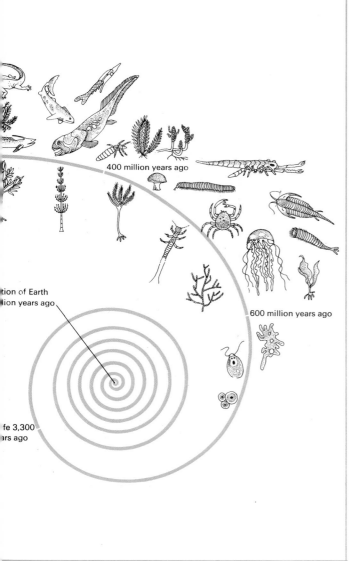

400 million years ago

...tion of Earth
...lion years ago

600 million years ago

...fe 3,300
...ars ago

ment, although paradoxically and interdependently it gave rise to challenge between the species. The most encouraging feature of contemporary thinking is the recognition of the total interdependency of all systems. We can no longer talk of the 'survival of the fittest', except as a truism, without taking into account the whole ecological energy budget and learning from the ecologists that a species can only survive if it has a symbiotic (organically co-operative) relationship with the rest of its ecological niche. To take one of Darwin's examples, the fastest wolf, which survives because of its greater killing ability, will itself face extinction if its prey is wiped out because it is too successful a hunter. That man is coming to realize that such laws apply equally to his own relationship with his environment represents a great evolution in consciousness. This point will be taken up in later chapters.

Oceans, Men and Exploration

FROM the earliest discovery that he could sit astride a log and float, man learned that he could travel as far as he could propel himself, or the wind and water would take him. From this simple beginning developed the history of boating, seafaring and finally the close scientific study of the oceans themselves – oceanography. And with this close study has come the realization of the oceans' significance in the destiny of the planet and of man, the self-conscious part of the planetary system.

Over 100 years ago, a U.S. naval officer by the name of Matthew Maury had the foresight to realize the tremendous value to mariners of a detailed knowledge of ocean currents. After spending a great deal of effort on charting the major global water movements, a second realization came to him. He had uncovered the subtleties of layered waters travelling worldwide at different speeds, at different depths, and often in quite opposite directions. The ocean currents, he wrote, are the arteries of the Earth. He saw them as the blood-stream of the planet, maintaining life and climates.

To extend Maury's analogy of arteries, the interchange of gases and vapours between the planet's water currents and airstreams can be paralleled to the process of breathing in the lungs of animals and men. Just as we need to breathe oxygen, which passes through the thin walls of the lungs and is carried by blood to all the tissues of the body, so sea life depends on the exchange of oxygen and water vapour across the sea surface, the interface with the atmosphere. The movements of the great waters created by the gyrations and turbulences of the air as winds, assisted by planetary spin, ensure that the oxygen-rich surface water reaches the deepest sea life. And, as will be shown in a later chapter, the pollutants that man is content to take into his lungs also parallel the radioactive fallout which is now distributed so effectively throughout the oceans by atmospheric distribution and exchange.

Man's body is over two-thirds water, and this proportion is remarkably similar to the fraction of the Earth's surface covered by the oceans. This link may be no more than an interesting coincidence, but the similarity in chemical constitution of sea water and blood is not considered coincidental by some serious biochemical opinion. It has been suggested that the near equal proportions of calcium, potassium and sodium that make up both man's body fluid and the ocean salinity are directly connected. It is taken as an inheritance from the time, hundreds of millions of years ago, when our remote ancestors had evolved from the single-celled to the multicellular system and adopted their own internal circulatory system. What is more natural than that

Man confronts sea where sea meets land. The ancestors of mankind emerged from the salty waters onto the shore hundreds of millions of years ago, and people show a remarkable inclination to return to the shore for summer relaxation every year. But the relationship with the sea is equivocal; love is blended with fear and respect.

the circulatory fluid should have been that at hand in the surrounding sea?

Before leaving the fascinating area of connections past and present with the global lifestream, attention must be drawn to the way in which early life forms inherited the flowing form that is characteristic of the behaviour of movement in water. A jellyfish creates mirror images of itself by pulsations to propel itself along. This exactly reflects the natural vortex-motion of moving water being injected into still water. This classical smoke ring or 'doughnut' pattern can be seen in ever-increasing complexities to underlie even higher complex animals like man. In topological terms, the anatomy of the human body resembles an elongated doughnut; the alimentary canal represents the hole in the doughnut. At the mouth or intake end, the environment is fed in; along the tube exchanges take place until finally the 'borrowed' material is returned.

Water, the unique fluid

What is the nature of water, a medium at once ideal for transport when calm and yet viciously dangerous when turbulent and storm-swept? Water's special quality has been beautifully described by Theodor Schwenk as 'sensitive chaos'. Its story begins with its structure.

The simple formula H_2O betrays nothing of the strength and shape of the bonds between the single oxygen atom and the two hydrogen atoms, from which so many of water's amazing properties seem to arise. Creating water,

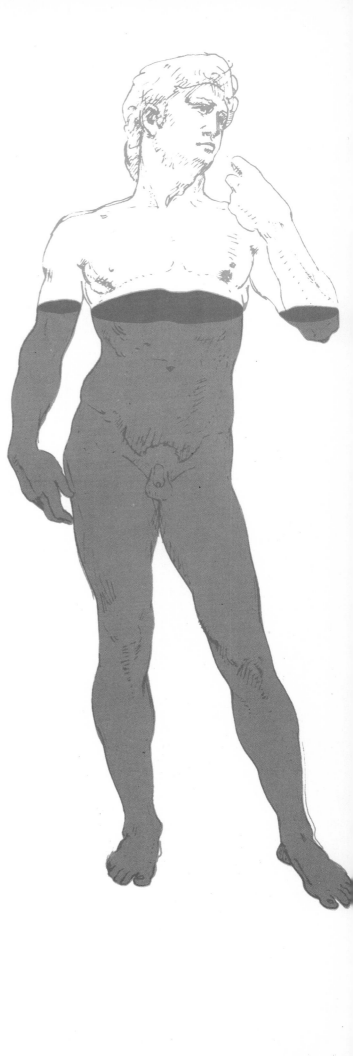

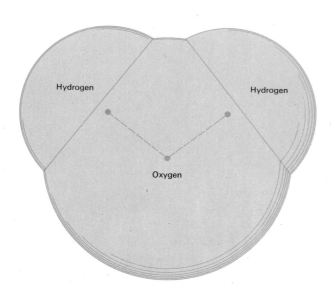

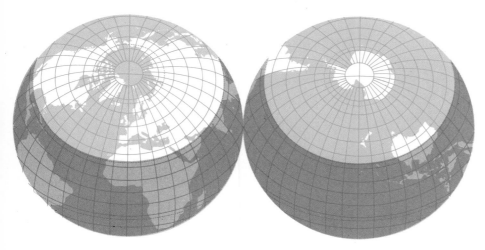

The affinity between life and the waters of the planet Earth is emphasized – although only coincidentally – by the mathematical proportions involved. **far left** The human body is more than two-thirds water; **left** a very similar proportion of the Earth's surface is covered by the oceans. The chemical make-up of the blood also echoes the oceans' saltiness, and this may be more than mere coincidence.
below The deceptively simple structure of the water molecule is responsible for water's amazing properties.

by joining hydrogen and oxygen atoms, is as easy as striking a match, but tremendous energy is needed to break the water molecule apart. This molecule is not symmetrical, but forms a dipole – that is, it has one end (the oxygen atom) negatively charged and the other (the two hydrogen atoms) positively charged. This lop-sidedness, with its peculiar effect on other chemical molecules, endows water with its properties as the 'universal solvent'. It is also responsible for the unique and – by the behaviour of other similar liquids – quite irrational property of being less dense at low temperatures than at higher ones. Due to the geometrical spacing of its molecules, ice sits irrationally on top of liquid water. If it behaved like all 'normal' substances, it would freeze from the bottom up. In cold weather, lakes and seas would rapidly become solid blocks of ice, for there would be no quickly-formed surface layer to insulate the deeper waters from further rapid cooling. The fact that ice floats means more for the global existence of life than almost any other of water's properties.

The great solvent powers of water mean that in oceanic form it can carry all the dissolved salts which enable it to remain liquid even well below the normal freezing point of 0°C. As the Ancient Chinese philosopher Lao Tzu put it, 'Water is good; it benefits all things and does not compete with them.' This colourless, odourless, tasteless substance maintains all the life forms by its very passivity. It brings in nourishment, carries away refuse and maintains temperatures, as well as acting as a universal lubricant during the whole process. Not only does water cool a fevered brow, but because of its remarkable capacity for absorbing heat, it maintains and distributes heat energy throughout the planet, having an enormous influence on climate.

Men can live for many weeks without solid food, but without water survival time is counted in days. Some life forms, notably certain bacteria, can survive without oxygen, but no creature can live without water. Plants, with a few exceptions, create their own nourishment from water and air with the aid of sunlight. And plants pull water upwards as sap – reversing the normal flow, as occurs so often in living processes. Plant-stems and tree-trunks act as vertical pipelines through which they suck up water and pass it out as vapour through the leaves in the process of transpiration. A single birch tree has been calculated to raise between 50 and 70 gallons into the atmosphere on a warm day. This unpredictable phenomenon has been described as one of the most intriguing puzzles of all biology: How can a sequoia, for example, pull up its lifestream sap 350 feet from the ground to the topmost branches? The answer seems to lie in the tenacious bonding between the water molecules, set in motion by the transpiration of water vapour from the leaves. This leaves a negative pressure which is followed up by the adjacent cells, with higher pressure. The sap thus climbs molecule by molecule, cell by cell, under the tension created by transpiration.

The vortex form of moving water is the basic shape of all but the simplest types of animal life. Even man represents an elongated but complex doughnut, the alimentary canal being the hole through the centre.

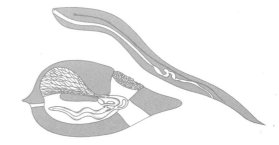

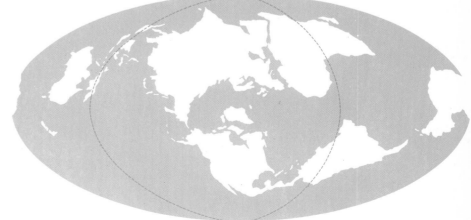

above Planet Earth or planet Ocean? The planet as seen from space by the Apollo astronauts. **right** This unusual projection of the globe shows the continents as islands in a great world-wide sea, demonstrating the fact that water covers more than twice as great an area as land.

As sap is the lifestream of a tree, so the movements of the oceanic waters are the lifestream of the planet. As already shown, water in all its states – ice, liquid and vapour – is a rare phenomenon which can occur in a mere 2 per cent band of the Solar System. Given full reign on planet Earth, water displays the properties that make it a unique vehicle for the emergence of life. The added property of salinity – resulting from the amazing solvent power of water and the constant erosion of the continental rocks by rain – modify even further those already amazing faculties. The salts in sea water have negligible effect on such physical properties as viscosity (stickiness), light absorption, heat capacity, or heat changes on freezing or melting, vaporization or condensation. But because of changes in density due to the amount of solids dissolved, saline water does increase the diffusion and spreading of these solids. As far as life is concerned, this adds greatly to the distribution of wealth. The phenomenon of salinity also results in osmosis (the ability of water to permeate a membrane from low concentration to high) and electrical conductivity – both properties favouring life processes. Finally, salinity increases the whole circulation of our collective lifestream and maintains the convection currents right down to freezing point. This too is responsible for an increase in nutrient distribution, sustaining the teeming oceanic life.

Into the undersea world

For all man's intimate long-term connections with the oceans, the most immediate facts of his confrontation with sea water are that he cannot drink it, it will not water his crops, and he can very easily drown in it. This in no way deters his fascination for the sea, as many sea-faring men who have risked their lives many times and returned will testify. At the bottom of the scale is the massive appetite industrially-developed, town-worn people have for even a day at the sea-shore in summertime. A more committed enthusiasm is shown by the popularity of underwater goggles, snorkels or skin-diving suits with breathing apparatus for the enthusiast – all inventions at the end of a long line of development by men who were driven by an all-absorbing ambition to enter the silent depths.

Where can one begin the story of man's fascination for the underwater? Certainly one can speculate that early man must have had to escape from a predator – and thereby discovered that he could swim. And fish as a tempting diet would undoubtedly have led him out into deeper waters and probably thereby stimulated the development of boats. With the mastery of water transport came the genesis of

sea power, trading and colonization. It is also fascinating to speculate on the point when man was tempted or driven into going *under* the sea's surface. There are dramatic legends of how Alexander the Great, 2,300 years ago, had himself lowered in a glass barrel supported by gold chains, in order to satisfy his curiosity about the undersea world. It is possible, though, that he was only repeating an everyday event in the work of those building his Mediterranean sea-port walls. Herodotus a century earlier spoke of marine science, and so did the Roman historian Pliny.

Leonardo da Vinci, as might be expected of such a multi-faceted Renaissance figure, designed underwater apparatus. The Venetian Buonaiuto Lorini in 1609 also designed a diving apparatus. A hundred years later, two probably eccentric Englishmen – John Williams in 1692 and John Lethbridge in 1715 – were actually lowered into the sea in a wooden case and a hogshead respectively. A colourful Frenchman, Abbé de la Chapelle, wrote a treatise on the diving suit which appeared in 1775. A German by the name of Klingert demonstrated great personal faith in his diving

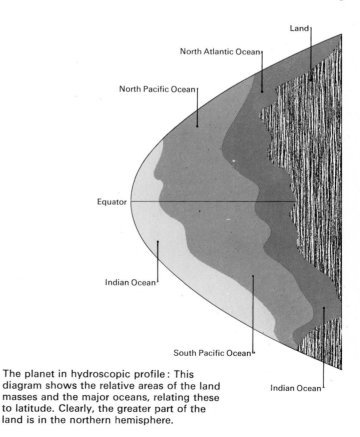

The planet in hydroscopic profile: This diagram shows the relative areas of the land masses and the major oceans, relating these to latitude. Clearly, the greater part of the land is in the northern hemisphere.

Phoenician galley: The Phoenicians were the first great ocean sailors of history, combining navigational skill (possibly derived from a knowledge of Babylonian astronomy) and seamanship. An Ancient Egyptian expedition probably manned by Phoenician sailors circumnavigated the continent of Africa over 2,500 years ago. Some anthropologists suggest that they even crossed the Atlantic to central America, bringing Mediterranean culture and knowledge to the American Indians of Peru and Mexico.

Norse longboat: This epitomized the conquest of the oceans by strength and courage – two qualities essential for successfully challenging the North Sea and North Atlantic. The vikings built slender craft, showing great economy in the use of materials. They really began the new era of European expansion and colonization after the Dark Ages, reaching Greenland and North America 500 years before Columbus.

Twin-hulled Tahitian canoe: The Polynesians of the Pacific Islands were skilled navigators, crossing thousands of miles of stormy seas and settling on islands from New Zealand to Hawaii. They found their way with navigational models made from sticks; where the sticks joined together represented an island.

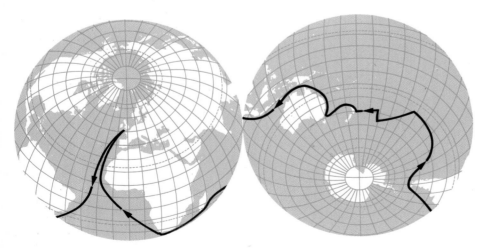

The history of Western civilization has been closely bound up with the history of shipping and seafaring, for until this century travel by ship was the only way of spanning any great part of the globe. **left** The voyages of Captain James Cook in the 18th century mark a turning point in navigation and the beginning of global man. Cook was the first scientist-navigator, and charts based on his surveys are still in use today.

Spanish galleon: The 15th and 16th centuries saw the rapid growth of European naval power and the start of the great era of exploration and colonization by European nations. The galleon was the principal type of ship used in this expansion. Men such as Christopher Columbus, Vasco da Gama and Ferdinand Magellan, sailing mainly out of Spanish and Portuguese ports, searched for new trade routes to the East and found new lands for Europe to exploit.

U.S. aircraft carrier: This represents shipping after the industrial revolution. The hull is of steel, the motive power is steam, generated either in a coal- or oil-burning boiler or in a nuclear reactor. Such craft are complete floating cities, communities that include all the specialists and tradesmen found in a city on land. They also represent the key to naval power in the era of global warfare.

Russian research ship *Cosmonaut Vladimir Komarov*: This space-tracking and research vessel is packed with radar equipment, computers and other electronic gadgetry. It represents shipping in the space age, the era of computers and automation following the second, miniaturizing industrial revolution.

Serious research into the oceans and their currents and the nature of the sea bed began in the 19th century, but primitive equipment made data-collection slow until the era of electronics in the middle of this century. Personal diving equipment, too, remained cumbersome until the 1940s, when the invention of the demand-regulated aqualung allowed long free dives (see pages 108 and 114).

The lead plummet was the sailor's only means of measuring the depth of water under his hull for many centuries. It was very difficult to use for depths of more than about 600 feet. A small hollow in the bottom of the lead weight could hold lard or some other greasy material, to pick up material from the bottom and show the nature of the sea bed.

The gun tube, invented by American scientist Charles Piggott in 1934, enabled samples of compacted deep silts to be brought from great depths. It contained an explosive charge which was fired automatically when the device touched the bottom. The result was to shoot the sample tube into the sea bed with great force. Samples of the Atlantic sea bed 10 feet long were obtained in this way.

The Ekman tube, invented by Swedish oceanographer Walfrid Ekman in 1905, enabled core samples of sea-bed silts up to 15 feet long to be brought up from the bottom. An arrangement of weights and ratchets was released when the tube struck solid bottom. The result was to close the bottom of the sample tube, keeping it enclosed to be brought safely to the surface.

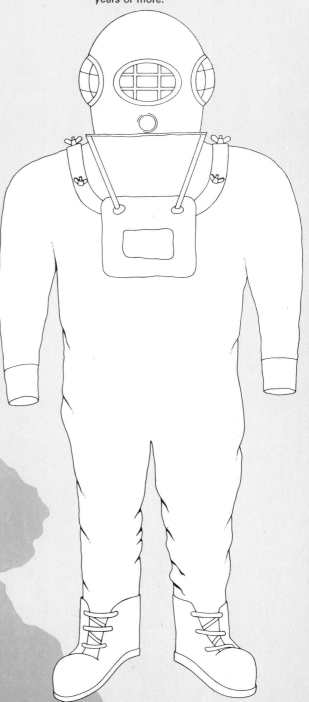

August Siebe's first 'closed' diving suit of 1837: It was not very different from the type used by naval divers for the next 100 years or more.

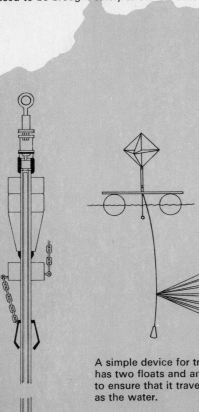

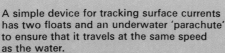

A simple device for tracking surface currents has two floats and an underwater 'parachute' to ensure that it travels at the same speed as the water.

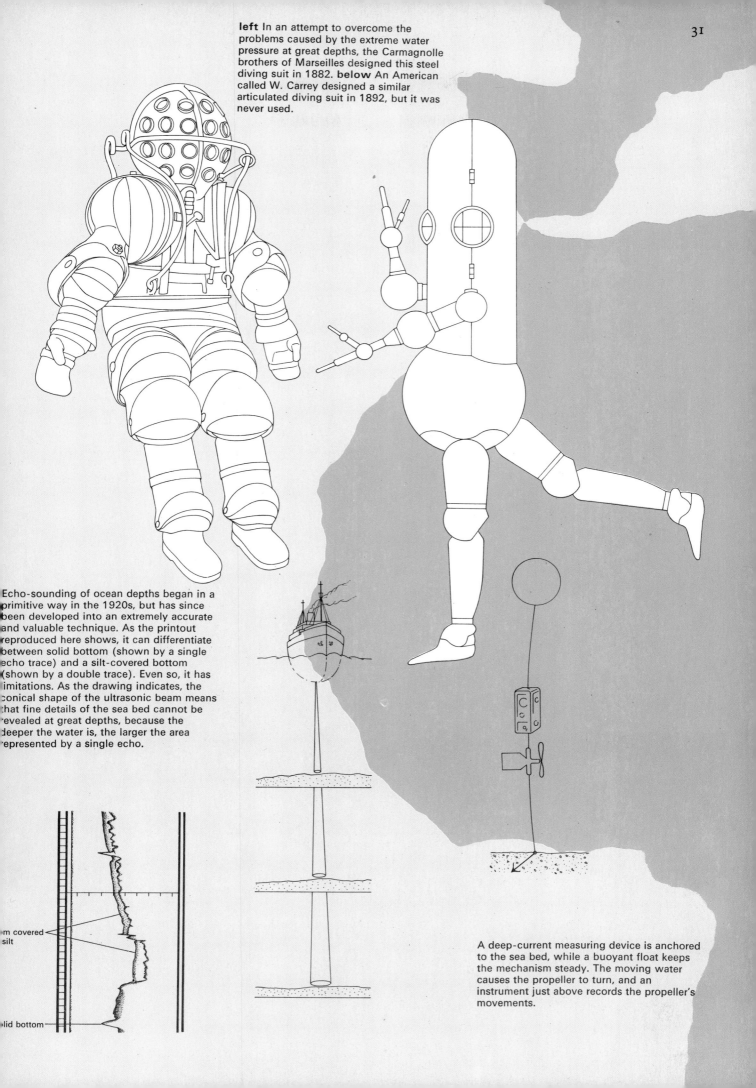

left In an attempt to overcome the problems caused by the extreme water pressure at great depths, the Carmagnolle brothers of Marseilles designed this steel diving suit in 1882. **below** An American called W. Carrey designed a similar articulated diving suit in 1892, but it was never used.

Echo-sounding of ocean depths began in a primitive way in the 1920s, but has since been developed into an extremely accurate and valuable technique. As the printout reproduced here shows, it can differentiate between solid bottom (shown by a single echo trace) and a silt-covered bottom (shown by a double trace). Even so, it has limitations. As the drawing indicates, the conical shape of the ultrasonic beam means that fine details of the sea bed cannot be revealed at great depths, because the deeper the water is, the larger the area represented by a single echo.

m covered silt

lid bottom

A deep-current measuring device is anchored to the sea bed, while a buoyant float keeps the mechanism steady. The moving water causes the propeller to turn, and an instrument just above records the propeller's movements.

The beehive-shaped *Hydrophilos*, designed in 1899 by L. de Rigaud but never built, was the fanciful forerunner of today's sophisticated, instrument-packed submersibles. The *Hydrophilos* had two main compartments joined by a central airshaft and spiral stairs. The lower compartment contained engines for driving the forward- and downward-pointing propellers. The hydraulic claw and grapnel are not very different from the manipulators of today's submersibles.

A variety of submersibles exist today, able to take oceanographers deep under the sea to study plant and animal life, ocean currents, and the sea bed. Those shown here include *Star III* (which can dive to 2,000 feet), *Deepstar IV* (4,000 feet), *Beaver IV* (2,000 feet) and *Deep Quest* (8,000 feet).
The *Aluminaut* is one of the larger present-day deep-water research submarines, with a length of 51 feet. Unlike most others, its hull is of aluminium. It is operated by a crew of two, and can carry four passengers to a depth of 15,000 feet. While they watch through small portholes, versatile manipulator arms can move heavy objects or probe the sea bed, transferring samples of silt or rock to boxes.

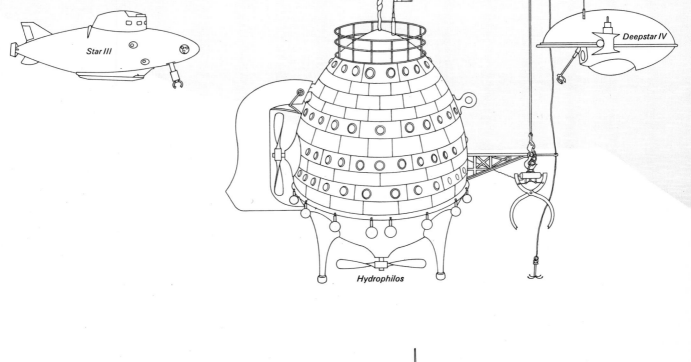

Star III

Deepstar IV

Hydrophilos

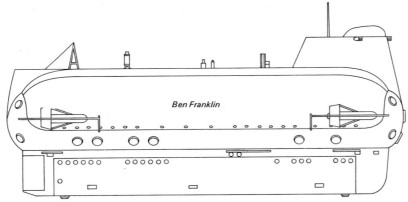

Ben Franklin

The Grumman-Piccard floating underwater laboratory *Ben Franklin* made its first voyage of discovery in July and August 1969. It drifted 1,440 miles 'down' the Gulf Stream of the Atlantic in 32 days. Led by Jacques Piccard, the team on board discovered that the Gulf Stream is far from the broad, steadily drifting 'river' it is commonly believed to be, but consists of 'several swirling, colliding, meandering torrents tumbling northwards'. The submersible was at one stage pushed right out of the Gulf Stream by huge eddies caused by hills on the sea bed.

The deep-ocean drilling ship *Glomar Challenger* provided the proof that the continents are drifting apart as some oceans widen and others contract. In the late 1960s and early 1970s, it carried out an extensive drilling programme for the Joint Oceanographic Institutions Deep Earth Sampling (JOIDES) group. It has drilled as much as 2,500 feet into the ocean floor in water several miles deep, bringing up core samples that confirmed theories of sea-bed spreading. The 140-foot derrick on the ship's deck performs the drilling operation, while thrusters in the bow and stern keep it steady. Hydrophones below the hull pick up ultrasonic signals from beacons positioned on the sea bed, and a computer analyses these and feeds control

signals to the thrusters so that the ship is kept in position over the drill hole.

Modern oceanographic research buoys can make many simultaneous measurements at various depths, and transmit information by telemetry to a data-collection centre. The buoy illustrated here is used by scientists working at Woods Hole Oceanographic Institution, and includes devices for measuring wind speed and direction as well as water temperature, pressure and current.

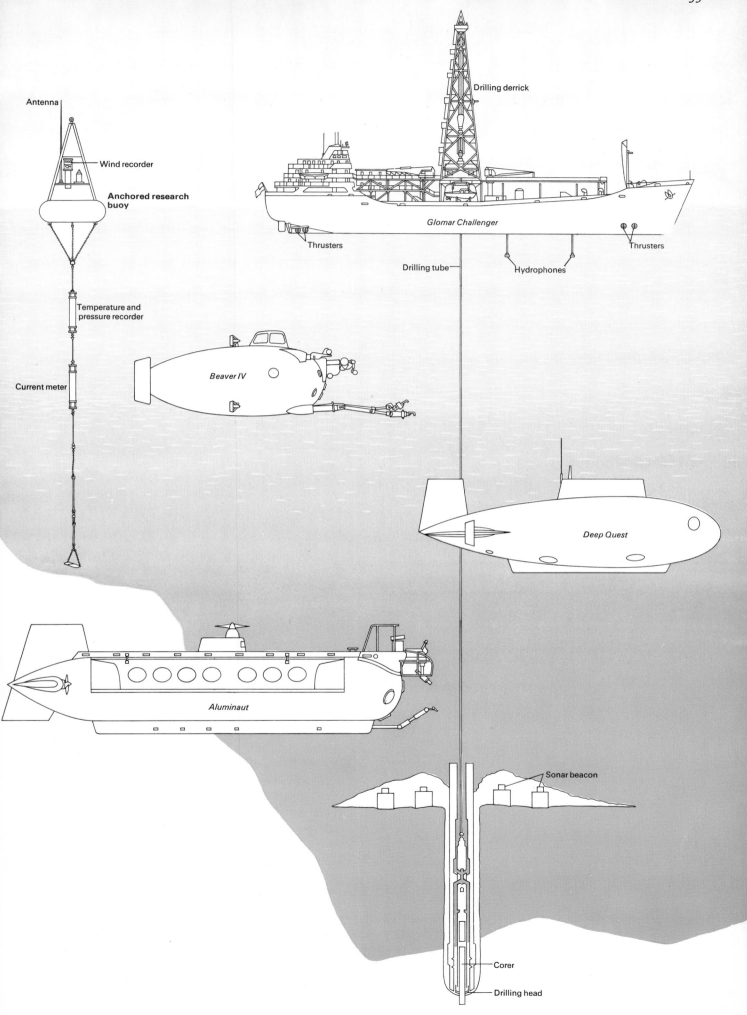

Antenna

Wind recorder

Anchored research buoy

Temperature and pressure recorder

Current meter

Drilling derrick

Glomar Challenger

Thrusters

Thrusters

Drilling tube

Hydrophones

Beaver IV

Deep Quest

Aluminaut

Sonar beacon

Corer

Drilling head

top Divers are lowered by crane into the sea beside a huge drilling rig. **centre** A satellite photograph of the Grand Bahama Bank, when analysed in the laboratory, can reveal water depths with great accuracy. **bottom** FLIP, or *Floating Instrument Platform*, is an oceanographic research vessel that cruises to its research station like a normal ship and then, by flooding ballast tanks, tilts to an upright position to provide a stable floating platform.

machine by descending into the River Oder in 1797. By 1860, the French navy were training divers and paying out underwater incentive money to tempt men into undersea service. The drive to get under the waves and stay there had really started. A quaint design by an American, Alvary Temple, in 1896 reflected the enthusiasm for the bicycle in the late 19th century; it was muscle-powered by pedals. Most ambitiously, Count Piatti del Pozzo built an impressive sphere and in 1898 descended 150 feet at Cherbourg in France. Another unusual American sailor, William Beebe, built his first bathysphere in 1930, and broke all depth records in the next four years.

In the 20th-century development of free diving techniques – for those wanting to roam at will in the depths – the important names are almost all Frenchmen. The one exception is a Russian by the name of A. Kamarenko. All were active on the French riviera from the 1920s onwards. Maxime Forjot and Kamarenko both developed the goggles, underwater breathing apparatus and underwater hunting gun. There followed a famous trio, all officers in the French navy: Jacques-Yves Cousteau, Philippe Tailliez and Frederic Dumas. These words of Tailliez sum up the fever of underwater enthusiasm: 'There was not a single ship of the Mediterranean fleet, from cruiser to despatch boat, in the years 1937–39, on which some junior officer was not constructing, with the complicity of the ship's workshop staff, an underwater gun designed to hunt grouper.' Although fish-hunting was the first incentive, the inventors of the diving equipment took a conservationist and anti-pollution stand as the years went by.

Commander Cousteau and his close friend Dumas were determined that they should all be able to stay below the sea for more than mere snatches of time. A historic meeting between Cousteau and engineer Emile Gagnan in Paris in 1942 led to one of the most important inventions for opening up the continental shelves for free-swimming men: the demand-regulated aqualung. This finally allowed long free dives, by equating the pressure of the breathing air with that of the sea and permitting the diver to inhale easily without having to fight the water pressure. In Cousteau's own words: 'To have enough time to explore the world of silence – that was what we had always longed for.' He had achieved it.

Into the age of high technology

Oceanography has come a very long way since the pioneer expedition of *H.M.S. Challenger* in 1872–76. Traditional oceanography of this type, where exploration and dis-

above left Jacques Cousteau's submersible *Soucoupe* was the first submersible to be in perfect equilibrium with the sea, needing no ballast. **above** *Sealab II* underwater habitat was used for an experiment off the Californian coast in 1969 until one diver died through an aqualung failure. **below** The Vickers submersible *Pisces* is seen with its support ship.

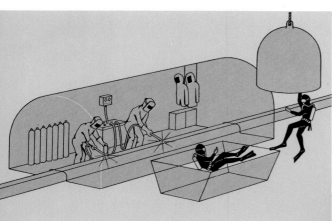

above SPID, an inflatable rubber saturation-diving habitat, was used in the second *Man in the Sea* experiment in 1968. Capable of being used at depths of 300 to 400 feet, it is built from a special rubber that does not allow helium gas to escape.

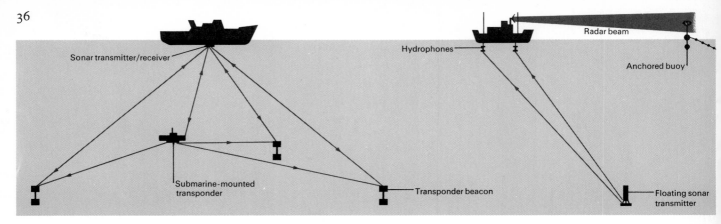

Sonar transmitter/receiver

Submarine-mounted transponder

Transponder beacon

Hydrophones

Radar beam

Anchored buoy

Floating sonar transmitter

covery are made from a ship sailing on the surface, still has an important role. Indeed, it was data on sea-bed magnetization recovered from such ships in the 1960s that finally confirmed the theory of ocean-floor spreading and continental drift. But scientific information on the oceans is now collected and recorded by many widely varied techniques. Satellites and aircraft can take infra-red photographs of the sea which show weather features, warm and cold currents, algae blooms, and many other features. Submarines and divers can work deep under the sea and observe ocean life in a way that was hitherto impossible.

There were two recent historic occasions that marked the birth of modern oceanology: the submerged crossing of the North Pole by the U.S. nuclear submarine *Nautilus* on 3 August 1958, and the descent to the bottom of Challenger Deep in the Marianas Trench by Jacques Piccard and Don Walsh in the bathyscaph *Trieste* on 23 January 1960. In 1965, Jacques-Yves Cousteau began an experiment which was to be the most dramatic breakthrough for diving since his development of the aqualung more than 20 years earlier – the first *Conshelf* experiment. Cousteau's team of divers lived underwater for one week at a depth of 30 feet, working from a cylindrical habitat moored on the sea floor and supplied with compressed air from the surface. The following year, Cousteau ran another *Conshelf* experiment; this time four divers remained submerged for a month at 30 feet, and a team of two at 80 feet for two weeks. Many other experiments have since been carried out on the possibilities of living and working under the sea, perhaps the best known being the American *Tektite* and *Sealab* programmes. The importance of surface activity has not changed, though, since most undersea experiments need a considerable amount of supporting hardware and personnel on the surface to monitor and service the undersea work.

More and more scientists feel the need to explore the oceans by diving and seeing for themselves, and archaeology, among other fields, has expanded into the undersea world and made many important discoveries there. But alongside these developments has come a new sophistication in scientific instruments and methods of measurement with remote-controlled sensors on the ocean bottom, ocean surface and at depths in between. With the additional information now available from surface and satellite monitors of ocean conditions all over the world, it is possible to make accurate predictions of the state of the sea. The importance of such systems for the utilization of ocean resources and as a 'thermometer' to check ocean health is immeasurable.

above Modern underwater sonar devices using transponders – receiver-transmitters that re-transmit a signal as soon as it is received – allow accurate positioning by triangulation. Sonar can also be used to plot deep ocean currents by tracking a float which maintains a certain depth.
right *Starfish House* was the underwater habitat used by Jacques Cousteau for his *Conshelf II* experiment in 1966.

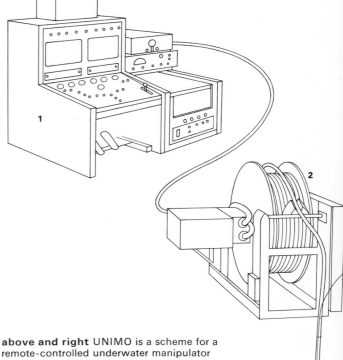

above and right UNIMO is a scheme for a remote-controlled underwater manipulator developed for the Shell oil company. It is designed for maintenance work on sea-bed oil well-heads at a depth of up to 1,000 feet, but could probably be developed for much greater depths. It is suspended by a cable from the mother ship's derrick, but has thrusters for adjusting and holding its underwater position. Powerful lights and television cameras give the operating crew a clear view of the job being performed, and this is itself carried out by versatile mechanical arms. Key to drawing: *1* Control console in mother ship, with sonar and television monitors. *2* Cable reel. *3* Floats to give buoyancy to bottom section of cable. *4* Power, control and television monitor cable. *5* Suspension cable. *6* Television cameras and lights. *7* Main propulsion unit. *8* Drift propulsion unit. *9* Manipulation arms.

below left As a route-map for the undersea voyage of discovery presented in this chapter, a new view of our planet has been devised. The projection has been specially computed to give a new, accurate and stimulating viewpoint on the oceans of planet Earth. The starting points for generating the projection are natural phenomena: four gravitational nodes, major anomalies in the planet's gravitational field discovered by an American satellite in 1954. By joining these points, the globe is divided by six boundary lines into four equal areas. The great-circle itinerary taken in this chapter follows a continuous route along four of these six lines, completely encompassing the Earth.

To show the route clearly, an octahedral projection using eight equilateral triangles has been computed, the gravitational nodes falling at the centres of alternate triangles. Our great-circle itinerary follows the lines that bisect these triangles. A copy of this projection accompanies each illustrated feature, indicating its position in the voyage and relation to the whole.

below Like astronauts taking their first close view of a planet, we look southwards, down into the Coral Sea and Tasman Sea from a position above the Solomon Islands. The eastern coast of Australia lies to the right, the twin islands of New Zealand to the left. This first glimpse of the undersea world highlights the dramatic relief of the sea bed in comparison with the eroded surface of the land. Islands are seen to be merely the exposed tops of undersea mountains.

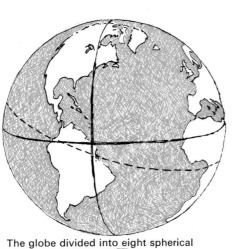

The globe divided into eight spherical triangles

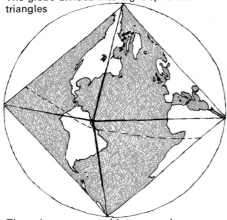

The sphere converted into a regular octahedron (eight-sided figure)

The unfolded net: an octahedral projection of our planet

Chapter 3

The Hidden Landscape

ALL the data that man's ingenuity, scientific skill and sheer bravery has won from below the sea-surface makes up a body of knowledge that speaks of undreamed-of topography, initially unbelievable behaviours, tensions and explosions that quite outweigh the kind of events that men experience on land. The ocean bed is the living skin of a living system – that is the new picture.

In order to explore this mysterious, hidden world, a special itinerary has been devised that sweeps around the globe on a direct route and by stages similar in length to the great bird-migration distances These jumps are along four great-circle edges of a four-fold division of the planet's surface (see diagram on the opposite page).

From carefully chosen vantage points on this itinerary, ten great features of the hidden landscape will be viewed, from close in or from the long perspective of space – but in each case with the thick covering of water drained away or made transparent. Other spectacles, too, will come under scrutiny as we journey around the world, but the ten features form a unifying framework; they represent topographical phenomena found repeatedly in the hidden underwater landscape. The ten features have been chosen not in order of significance, but simply in order of occurrence as the route unfolds. In order of appearance, these are: the great mid-ocean ridge system, the continental rises, the asymmetric ridges, the continental shelves, the deep trenches, the oceanic volcanoes and island arcs, the continental slopes, the canyons that cut through these, the abyssal plains, and finally the oceanic rises.

Until the visual display-instrumentation and processing of an immense amount of data is forthcoming, the best way of presenting these beautiful and magnificent phenomena is through the artist's vision. The facts are based on the best available scientific data brought alive by the imagination of the artist. Each feature as it is displayed will be discussed in terms of its function, size and behaviour, according to the most up-to-date information and theory. Naturally, as in all healthy sciences, there are opposing theories. This reflects the rejuvenating and revolutionary process that the whole of geological science has been experiencing since the mid-1960s.

The great-circle itinerary demonstrates time after time how man has yet to see and fully comprehend these amazing aspects of his planetary home. Modern geology is experiencing the same kind of revolution as modern physics, with dynamic concepts replacing static ones right at the heart of what we thought was so stable, the ground we stand on.

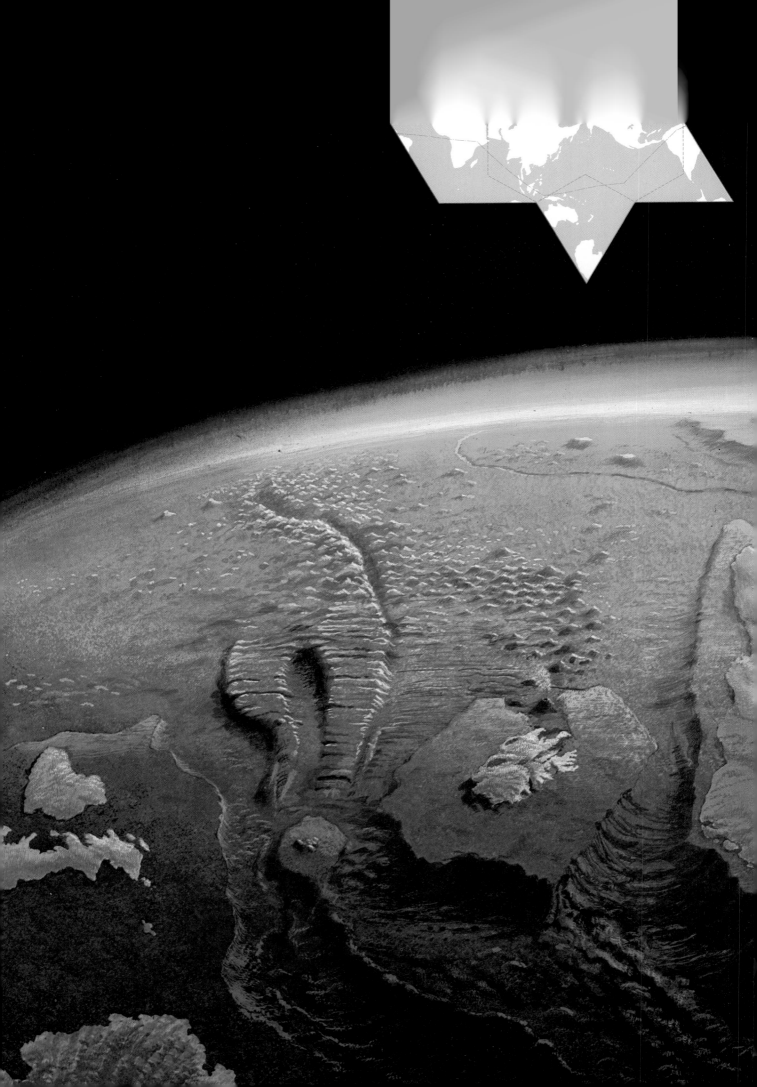

Mid-ocean ridges

The central ridge of the ocean floor is like a primitive streak, the spinal origin of the embryonic sea beds. Current theory says that it represents both the original division of the continental land masses and the growth-centre of sea beds. Because this massive mountain range forms a seam running through all the world's oceans, similar to the join in a tennis ball, it is the starting point of our great-circle itinerary. It is shown here in the North Atlantic to the south of Iceland.

The very scale of this planet-encompassing mountain range alone is cause enough to reassess the concept of mountains, which on land have had such a determining role in human civilizations. But there is another difference; the undersea ranges have experienced virtually no erosion and are as starkly pristine as they were born. The only addition is the constant 'snowfall' that characterizes the sedimentary ocean environment.

Another difference is in construction; land ranges are built up in compression, those of the mid-ocean in tension. The nature of sea bed production, by spreading on each side of the great seam, embodies a unique form in the crest of the range. Instead of a series of climax peaks, there is a gigantic rift valley far out-spanning the breath-taking Grand Canyon. The peaks rise to 6,000 feet on each side of this, on average, 30-mile-wide fissure.

One end of the 40,000-mile-long mid-ocean ridge system can be considered to be in the Laptev Sea off the Arctic coast of Asia; the other, having traversed all the major oceans, disappears under Alaska. We have yet to discover if there is a form of linkage between these two terminals. In some places on its serpentine course, the ridge has reared so high as to burst out into the atmosphere and form islands. Iceland is one such surface-breaking drama, with its active volcanoes and hot springs; here the rift valley is displayed in broad daylight. Here, too, the spreading to either side of the ridge can be measured directly.

Following the ridge down the Atlantic, we find it emerging a number of times as the islands of the Azores, Ascension, St Helena, Tristan da Cunha and Bouvet Island. On reaching the South Atlantic, the ridge rounds the Horn of Africa only to divide. One branch continues right up to the Gulf of Aden. The other extends south-east to divide Australia from Antarctica, and sweeps right across the South Pacific to divide at Easter Island. A small branch heads for Chile while the main line shudders with huge fault lines up to California and on through to Alaska.

Our knowledge of the mid-ocean ridges is incomplete, and some fascinating puzzles remain. Are the gulfs of California and Aden new-born oceans, as the seams suggest? Where do the American and Eurasian plates meet? The questions are large, the answers yet to come.

The six major crustal plates of the Earth: This map shows the surface opened out flat to indicate the relationship of each plate to the others. The light-toned boundaries show where new crust is being formed and is emerging from the mid-ocean ridges. The dark-toned boundaries indicate plate edges where material is being swallowed into the mantle or is being crushed into mountain ranges. In fact, geophysicists have identified about a dozen other minor plates between parts of the major plates shown here.

Pacific plate

American plate

Eurasian plate

African plate

Antarctic plate

Indian plate

42

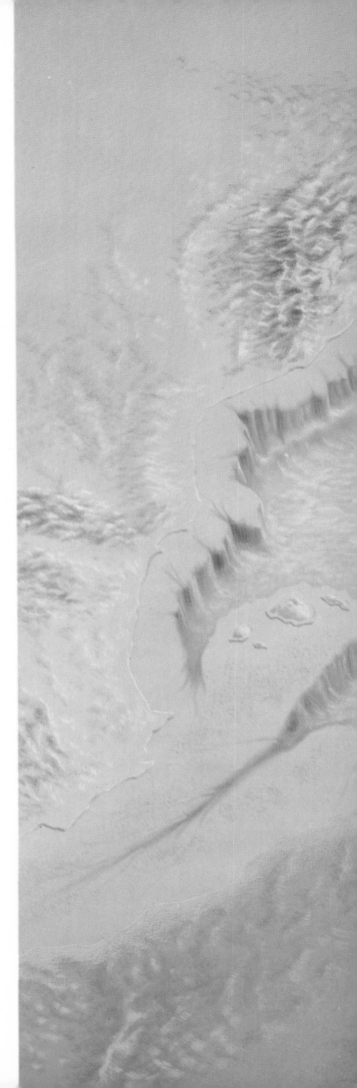

bottom A three-dimensional diagram shows salinity gradients and currents at the Straits of Gibraltar. Great evaporation of water from the surface of the Mediterranean creates dense, salty water. This leaks out into the Atlantic below the inflowing less-salty waters. Relatively little exchange takes place, however, because of the narrowness of the opening.

The Mediterranean basin

Both the bottom and the life of a landlocked sea are bound for many reasons to be quite different in character from the larger open ocean. Just as the ecology of an island is not so luxuriant, complex and resilient as that of a continent, so an 'island' sea surrounded by continent does not have the constitutional variety that makes for the splendrous marine life of the oceans.

In particular the changes of density and salinity and the diffusions that bring up nutrients from the bottom are very much less in evidence. Both the mixtures of temperature that characterize the Atlantic circulations and the Coriolis effect that helps spin the greater seas are absent or ineffective in the smaller Mediterranean.

Most of the Mediterranean is quite shallow, and had the water level remained during the last 4,000 years as it was in the Ice Age, thus linking with land bridges many centres of ancient civilization, the changes in Mediterranean history would have been incalculably different. This paradoxical nature of the sea – the medium that both separates and unites peoples – is brought home by the illustration of a waterless Mediterranean on these pages.

Today, this contradictory nature of water is emphasized in a different way. It is both man's drain and his water supply. And the modern Mediterranean is the most polluted of seas. Is it heading for a lifeless future like the Great Lakes of North America?

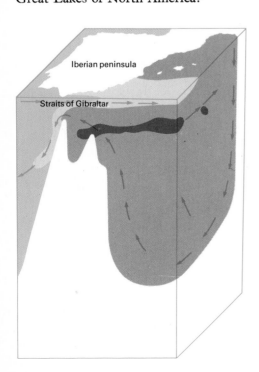

Iberian peninsula

Straits of Gibraltar

Continental shelf
feet
0
Continental slope
Upper continental rise
Lower continental rise
Abyssal plain
10,000
Oceanic rise
20,000
0
200
400
nautical miles

inset A profile of the continental slope and continental rise (with the vertical scale exaggerated) shows the distinct upper and lower parts of the rise. Current terminology insists on giving the name 'slope' to massive features with the steepness of cliffs.

Continental rises

Our next port of call allows us to focus in on the second major feature of the undersea landscape. This is at the foot of the massive edge of the African continent off Somalia. The continental rise is the very particular, gently-sloping surface that links the bases of these immense cliffs, the continental slopes (which are the true edges of the continents) to the deep ocean basins.

We have yet to fully understand the forces behind these long gentle slopes, which may vary in width from under a hundred to several hundred miles, and extend in length for thousands of miles. A double build-up of sediments seems the likely cause. On the one hand, the rises represent the area of the oldest sea bed as it spreads from the mid-ocean ridge. Therefore, in theory, they have the thickest covering of sediment. On the other hand, they represent the area which receives the outfall of sediment slurrying down the steep continental slopes.

The nearest land equivalents, on a much smaller scale, are features like solifluction or saturated soil flow down hill. Another similar topological parallel is the type of rockslide often known as screes. One huge such earthslide is Slumgullion Gulgh in the San Juan range of Colorado, U.S.A., which is six miles in length. However, we must beware of land comparisons as the conditions are very different, however similar the form appears outwardly.

The continental rises, whose depths range from 4,500 to 17,000 feet, are divided by oceanographers into upper and lower regions, but both are essentially terrace- or shelf-like features. Those of the Atlantic are the best known, and from studies of these their general characteristics have emerged. The demarcation between the dramatic cliff-like climb of the slope and the upper continental rise is in some places sharply defined and at others has a transition area of up to six miles. The interesting subtleties of the rises were charted by American oceanographer Bruce Heezen in 1959, and show a remarkable repetition in the profiles. Both upper and lower rises subdivide into three segments, with a very gently-sloping middle area but a rather steeper slope above and below. Overall, the gradients vary between 1 in 50 and 1 in 2,000.

Occasionally the relatively uniform and persistent slope of the continental rises is relieved by hills 150 to 600 feet high and each from one to three miles wide. Otherwise those continental slope canyons that extend into the rise areas are the only abrupt features. The seaward boundary to the rise is usually clearly defined as it reaches the flat abyssal plains or oceanic basins.

Asymmetric and aseismic ridges

The next vantage point on our itinerary is the second tetrahedral node within the octahedral projection. From an elevated position, a very long perspective looks north-west towards the Bay of Bengal, with India to the left and Sumatra and Java to the right. Between them lies what rates as geometrically the most unlikely phenomenon on the planet: a virtually straight ridge extending 2,500 miles. It is like a huge finger pointing south towards Antarctica.

This asymmetric ridge has no comparison anywhere on the globe. (The nearest approach in straightness is the southern half of the Andes.) Known as the East Indian Rise or the Ninety-East Ridge, it is shrouded in mystery, because its origins and function in the whole geological picture is so uncertain. Could it be the track of a migrating continent? Geologically, it is categorized as an asymmetric ridge because it cuts virtually diagonally across the Indian plate between the active mid-ocean ridge to the west and the limits of the Java Trench to the east.

The East Indian Rise cannot quite be included among the aseismic (earthquake-free) ridges, since a few earthquake epicentres have been recorded at about the halfway point. There are, however, four other aseismic ridges in the Indian Ocean – the Mascarene Ridge, on which the Seychelles sit; the Maldive Ridge, which runs down from India and is crowned by the Maldive and Chargos Islands; a southern extension of Madagascar; and the southerly ridge on which ride the Kerguelen Islands.

Several similar ridges are found in the Atlantic, but the most surprising feature here is a huge transverse ridge at right-angles to the Mid-Atlantic Ridge. It runs from Greenland, through Iceland to the Shetlands, and southwards parallel to the British Isles, where it is known as the Rockall Plateau. A further ridge lies under the north polar cap, but they are far less frequent in the Pacific. Apart from three relatively small ones off South America, there is only the dignified underwater mass off the eastern seaboard of New Zealand, the Chatham Rise and New Zealand Plateau.

This whole issue of intermediate ridges – those apart from the mid-ocean ridge system – cannot be left without mentioning how little we know about them: their cause and function in the formation of ocean beds; their details and history. During the Indian Ocean International Expedition of 1959 to 1965, survey ships crossed the East Indian Rise 30-odd times, but not one survey was taken along its length. No wonder it is among the least known of major ocean features.

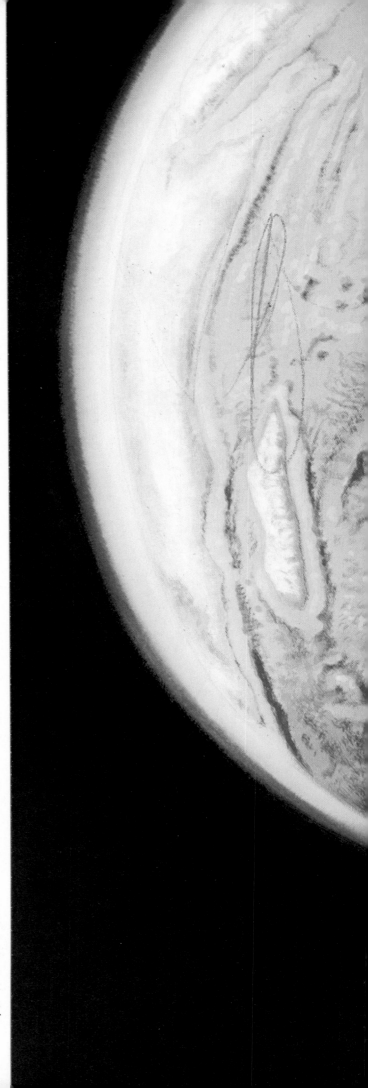

Continental shelves

These shallow extensions of the dry continents are for mankind the first, the last, and the most relevant part of the undersea landscape. They are first in as much as it was imperative for the navigators of history to fathom and chart their ways in and out of port, thus ensuring that valuable cargoes which had been nursed and buffeted across the world's oceans were not ignobly sunk on the home rocks of the harbour sound. They are last in as much as they are the latest areas of exploitation by man, in both the positive and negative sense. They are also the latest to be included in the land-claiming habits of ownership-bent humanity.

They are most relevant in terms of human habitability and future mariculture and marine farming – *if* we can radically alter our polluting habits.

The next focal point on our great-circle itinerary of the planet is the Arafura Sea, an area of continental shelf which, together with its neighbours the Gulf of Carpentaria and the Timor Sea, almost equals in area the entire West European shelf, including the British Isles. Bounding the Arafura shelf area to the north is the island of New Guinea, with its mountainous backbone but swampy southern shore, while to the south lies the Australian coast. The contrast between the relative flatness of this shelf and the tumultous contours of the neighbouring Banda Sea is dramatically emphasized by the 24,400-foot drop into the Weber Deep.

The southern side of the shelf covered by the Timor Sea has the major port of Darwin, and is oceanographically categorized as 'moderately known', whereas all the rest is classed as 'poorly known'. (This does not disguise the oil exploration all around Cape York Peninsula and throughout the Timor Sea.) But this shelf and other major shelves in the area have been assigned as areas of high biotic potential by Norwegian oceanographer Harald Sverdrup, once more highlighting the conflict of meaning in the term 'resources'. Under international control, mariculture in this area could bring great relief from the food shortages of an area of the world with particularly high population pressures – a subject which is expanded in a later chapter.

The continental shelves can be compared in many ways to the land they so closely relate to. Plant life abounds in comparison to the greater depths because of the amount of filtered sunlight available for photosynthesis. Seaweeds attached to the rock-strewn bottom act as ecological niches to many forms of animal and fish life – as well as forming a valuable source of food and drugs for man. The shallow waters are highly susceptible to both lunar (tidal) and solar (warming) influences, with resultant increases in circulation and diffusions – another factor favourable for sea life.

A further similarity with the land is in the silt-laden bed upon which so much of the life depends. These silts are the washed-away soils from land, and here they are first deposited before they are consolidated, or turbidity storms take them on down the continental slopes to the abyssal plains. The silts are rich in phosphates, silica, nitrates, calcium and other minerals which are needed by all oceanic life. (The Carpentaria Gulf in this instance has ten silt-rich rivers draining into it.) It is not surprising that 90 per cent of the human world's marine food comes from the continental shelves.

In total, the world's continental shelves represent an area roughly equivalent to all Europe plus South America. Their maximum depth below sea-surface level varies from about 400 feet to the deepest at between 1,200 and 1,800 feet. Their widths – controlled by long-term geological behaviours – vary between 20 miles, the average in the North American continent, to 750 miles across under the Barents Sea.

Their most dramatic period in recent geological history was when the waters were relentlessly drawn up into the massive glaciers of the Pleistocene Period. As a result, the sea levels of the world dropped some 500 feet, and the shelves emerged as part of each continent. Four times, in the four successive Ice Ages, this is known to have happened.

Once open to the air, the invasion by plant and animal life from dry land began. Off the eastern coast of North America evidence has been salvaged of roaming musk oxen, mammoths, horses, tapirs, giant sloths and giant moose as well as luxuriant forests of pine, oak and other trees. There have been similar finds off Europe and eastern Asia, where land linked Britain with France, Holland and Denmark, and linked Japan with China and Korea. Important human migrations are also believed to have accompanied the last of these periods.

Surveying the continental shelves of the world, the most outstanding fact is our lack of knowledge. It is almost paradoxical that we have satellites orbiting the planet that can, through infra-red photography, tell the state of every crop in every field, and yet we know so little about the near-three-quarters of the planet covered in water. It is particularly ironic that the vast life-rich areas of shallow seas off our coasts should, in spite of their proximity and long history of exploitation, be so poorly known.

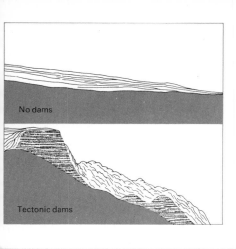

No dams

Tectonic dams

Continental shelves may be formed in various ways. **left top** They may result simply from the settling of sediments over millions of years, with no upward projection of rock to hold them in place. Such shelves have a steady, gentle slope. Or there may be rock formations behind which the sediments build up like water in a reservoir.

left bottom If such rock formations are formed by lava or by the uplift or folding of the Earth's crust, they are termed tectonic dams. **right top** Diapir dams result from the upward pushing of salt domes, underground salt deposits. **right below** Sediments may also collect behind reef dams, formed by marine organisms.

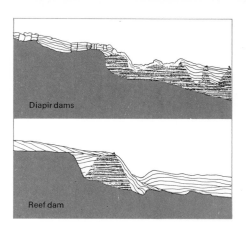

Diapir dams

Reef dam

Deep trenches

The deep trenches are currently believed to be the planetary incinerators – but on a time-scale that almost outlives the whole of human history. Here the plates that are created under tension along the mid-ocean ridges meet their useful end as oceanic bottom and are apparently forced under the lighter basaltic rocks of the continental mass that confronts them. As they go down – with the accompaniment of quakes and volcanic outbursts – they are believed to take sediment and refuse down with them for recycling within the molten circulations of the mantle's convection currents.

It is an amazing picture: Billions of tons of what man has always experienced as rock-steady solid earth are actually, on another time-scale, as much in circulation as the waters and atmosphere. Maybe it serves to emphasize the realization that the law of being on Earth is the law of circulation, of cycle and recycle, of birth, growth, maturity, decay and death.

The depths of the great trenches reflect the relative sizes of the oceans. Of the 20 greatest both in length and in depth, 15 are in the Pacific Ocean, and the first seven in order of magnitude are also Pacific. Only in eighth position comes the first and deepest Atlantic trench.

Philippines Trench	37,800 feet deep	750 miles long
Marianas Trench	36,200 feet deep	1,250 miles long
Tonga Trench	35,700 feet deep	780 miles long
Kuril-Kamchatka Trench	34,600 feet deep	1,340 miles long
Japan Trench	34,000 feet deep	940 miles long
Kermadec Trench	32,950 feet deep	750 miles long
Bonin Trench	32,150 feet deep	310 miles long
Puerto Rico Trench	30,150 feet deep	500 miles long

The deep, dark trench visited here is the third in this list: the Tonga Trench. It cuts an enormous gash through the south-western Pacific almost along the International Date Line, and acts as an invisible separation line between different Polynesian peoples. The Tongan Islands form a kind of vanguard to the Australian continent, forcing the South Pacific ocean bed beneath them as it advances.

Little is known about these forbidding and immensely pressurized areas (water pressure reaches 8 tons per square inch), but mention must be made of Lieutenant Don Walsh and Jacques Piccard (son of the famous Auguste Piccard). The two Piccards together designed the bathyscaph *Trieste* which took Walsh and Jacques Piccard down a historic 35,800 feet in the Marianas Trench on 23 January 1960. They became the first truly penetrating Inner Spacemen.

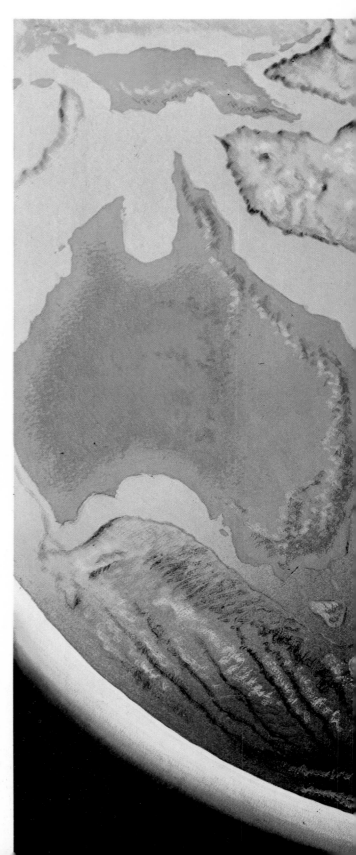

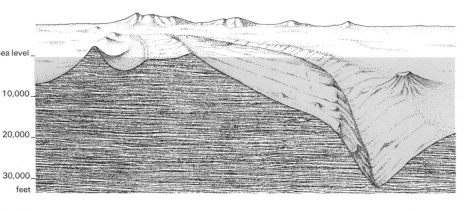

Sea level

10,000

20,000

30,000
feet

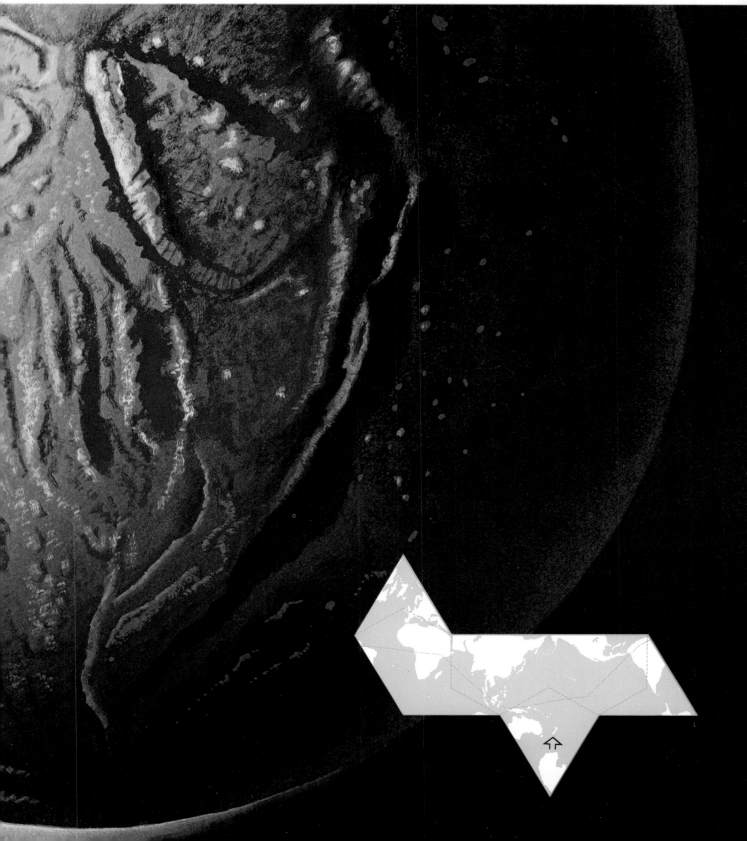

Oceanic volcanoes and island arcs

From the view of the Tonga Trench looking down from the Caroline Islands, the next stop on our journey is at the third tetrahedral node. From this vantage point deep in the Pacific region we look across to a dramatic row of volcanic peaks: the Hawaiian Islands.

To get some idea of the power, persistency and strange individuality of these volcanic islands, pause for a moment to think of the energy involved in one eruption. In the 1815 eruption of Tambora (in Indonesia), energy equivalent to 4,200 hydrogen bombs was released – in one event. Next to this, bear in mind the immense pressures of the water above an aspiring volcanic island. Up to a depth of 8,000 feet, a volcanic explosion is completely subdued by the oceans: the gases dissolve into the water and the lava spreads and hardens on the sea floor. Finally, put the fact that from the sea bed to the twin peaks of Mauna Kea and Mauna Loa (on Hawaii Island) is 30,000 feet, the last 13,800 feet towering majestically above the present level of the Pacific Ocean.

American marine biologist Rachel Carson suggested that this 2,000-mile-long range of volcanic islands, which arose during the Cretaceous Period, may have more than a coincidental relationship with the fact that the world was experiencing the greatest floods in its history. The single volcanic rise of Bermuda has been calculated to involve 2,500 cubic miles of volcanic matter. No wonder the displacement of the 2,000-mile Hawaiian chain of such rises figures in the calculations of changes in water levels over the whole globe.

These islands are just a few of the many volcanic cones that have managed not only to break surface but also to maintain resistance to the eroding forces of the waves and atmospheric elements. It has been calculated that off the coast of California alone there is an underwater volcano over 3,000 feet high on average every 25 miles.

Certain other characteristics have been observed about the connection between volcanic island arcs and frequency of actual volcanic occurrence. The greater the curvature of the arc, the more live cones there are – hence the implication that the more intense the crustal deformation the greater is the likelihood of plutonic material being erupted. The island arcs down the huge western (Asiatic) coast of the Pacific, with backs curved against the incoming plate that plunges back into the hellfires of the mantle just before them, seem to be related to the pressure of the diving plate squeezing out molten rock as multiple safety valves to the whole dramatic operation.

Continental slopes

One candidate for oceanographic renaming must surely be
the continental slopes. If ever an escarpment or cliff face
had a better case for being so named it is this, the most
consistently incisive feature of the whole planet. The word
slope carries a gentle connotation, but here we find an
incline that descends at the treacherous pitch of 2 in 1 for
almost twice the height of the southern wall of the Himalayas! These decisive banks have been called the true
edges of the continents – a fact clearly brought out during
the Ice Ages, when the continental shelves became dry
land and the ocean lashed the cliff edge of the steep
continental slopes.

How did demarcations of such insistence come about?
Current theory places the continents, made up of sial
(light rock), in a position similar to icebergs floating on the
sima, or more dense rock, of the crust. The edges of the
lighter rock are believed to correspond to the continental
slopes. It is often here that the ocean-bed plates meet the
bulk of the continents, and cause folding and buckling in
the continental edge and the plunging of the bed-plate
back into the crust-melting convection currents of the
mantle.

Over the millions of years that sediments have been
flushed down valleys from the upper reaches of the rocky
continents, a great variety of fragments and silts have been
deposited at the edge of the sea line. One could describe
the massive continental slope escarpments as the angle of
accumulation that characterizes this continuous dumping.

The artist's image on these pages focuses on one of the
most emphatic and stark of the slope regions: one where the
cliff-like edge to the South American continent descends
not to meet the gradual incline of a continental rise but
continues down into the equally stark and forbidding
depths of one of the planet's deepest fissures, the deep
trench of Chile. To put this drop into perspective, compare
the highest point on the South American continent, Mount
Aconcagua, towering 22,835 feet above sea level. The
descent down the continental slope from sea level into the
deepest part of the trench is 26,400 feet – nearly 4,000 feet
greater than the height of Aconcagua.

Biologically, the descent is from the green pastures of
the algae-rich sunlit zones with familiar food chains down
into a plant-barren world of carnivores preying upon each
other and scavenging the falling bodies of the inhabitants
of the lighter waters. The continental slopes see the gradual
transition of light, life and land to darkness, death and
depth.

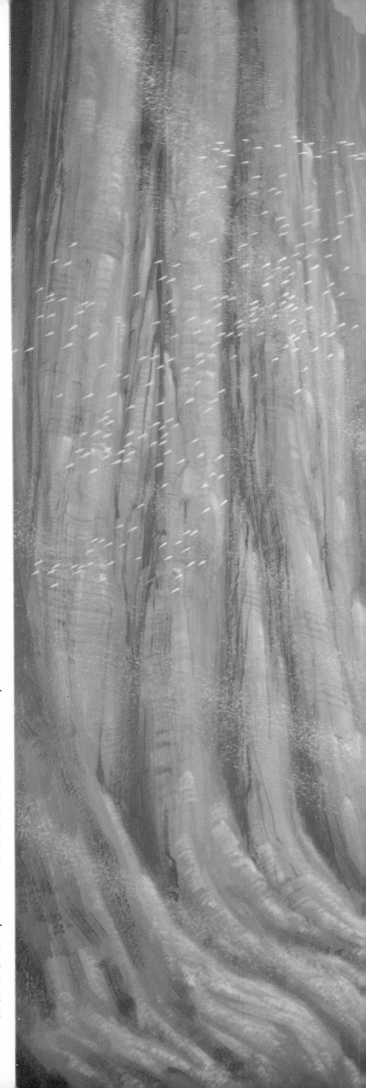

Continental slope canyons

From the fourth node, deep in the richest anchovy fishing grounds of the world, we will take our one major detour from the strict route along the tetrahedral edges to visit the intensely interesting part of the globe where oceanic creation, land mass movement and human destiny are so closely interwoven. This detour is to that proto-ocean, the Gulf of California, and beyond to Monterey Bay – site of a magnificent, complex continental slope canyon.

Monterey Bay is nearly 30 miles across, and at present two medium-sized rivers drain into it. But only one of these seems to have a direct bearing on the three major gorges of the submarine canyon in the slope to seaward.

The cause of such canyons must rate as the most hotly disputed subject of all oceanography after the continental drift theory. For at least 50 years they have been debated and argued over, with orthodox opinion holding that they were river beds of glacial times (when sea-levels were lowered). Then in 1936 a brilliant young geologist at Harvard, R. A. Daly, proposed the radical view that the canyons had been gouged by turbidity currents of the kind described in principle in the account of the abyssal plains later in this chapter.

Daly visualized the beginnings of these sediment-laden currents in times of lower sea levels, when much unconsolidated sediment of the shelf and slope was attacked and saturated by heavy rains and violent storm waves. In recent history, the tragedy of Aberfan, in South Wales, was the result of a similar principle: A silt hillside lying at its own angle of settlement became saturated from an internal spring, and a massive mudslide engulfed a village school.

To return to Monterey Bay, the fact that of the three heads of the canyon, only one could be directly related to a river mouth reinforces Daly's arguments against the ancient river bed theory. In fact, the turbidity could itself be caused by the river flow, but this would have to be particularly dense. Quite different beginnings seem to be indicated by the Carmel Canyon branch of the Monterey group. This dramatic feature drops from the 300-foot level to 6,500 feet over only 16 miles, exhibiting the immense power, mass and velocity of the silt-laden turbidity currents that carve out millions of tons of continental wall on their route to the lower levels.

The Congo Canyon in the West African continental shelf has amazing cliff walls 2,000 feet high on each side. Another better-known canyon is the well-documented Hudson Canyon off New York City, which carves its way well into the continental rise as well as the slope.

Transform fault lines

The Earth's solid crust is only one-hundredth the thickness of the fluid, moving mantle beneath it, so it is not surprising that the rigid surface sometimes has to form cracks to adjust to the constant movement. The world-wide network of transform faults associated with the mid-ocean ridges represent this adjustment. The surprising thing is that they are not far more uneven. Indeed, modern oceanographers are delighted that such extensive phenomena can be predicted by a relatively simple mathematical model of the globe – one that has existed, in fact, for 200 years.

British geophysicist Sir Edward Bullard has even gone so far as to suggest that marine geology is simpler than its continental counterpart, and that this is not an illusion based on our lesser knowledge of the oceans. He was particularly concerned with the spidery fault lines – often thousands of miles long – stretching from the Californian coast to the Hawaiian Islands, which are seen clearly in the artist's image on the opposite page.

These fracture zones, where huge chunks of crustal rock have slipped past each other, are directly associated with the centres of ocean-floor spreading, the mid-ocean ridges. However, their directions are not governed by the ridges, but by the geometry of a spherical globe. These principles were first propounded in the 18th century by Leonhard Euler, a mathematician friend of Beethoven.

Euler demonstrated that where spreading between two rigid plates takes place on the surface of a sphere (as is now known to occur on the Earth's surface), then the relative movement of these plates must be a rotation around a certain point on that sphere. This point becomes one pole of an axis of spreading passing through the centre of the sphere, and sliding movement between two crustal plates should be at right-angles to this axis – that is, along latitude lines of the axis (see diagram). A second prediction of Euler's model is that the rate of spread should be greatest at the 'equator' of spreading. Bullard's surprise was at how often these principles operate on the sea bed.

Thus the transform faults represent an adjustment the crust has to make to the spreading movements of the ocean-bed seams, the mid-ocean ridges. Oceanographers have also been surprised that the mid-ocean ridge develops new sea bed equally on each side in spite of the obstruction of a continental mass. One would expect the rate of ocean spread to 'close up' on one side. The reason it does not seems to be that the producing ridge will move away from resistance quite flexibly in order to maintain its balanced output to either side.

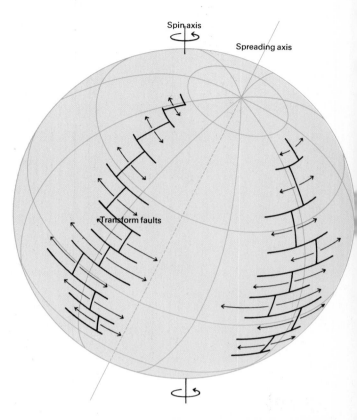

above When rigid plates move in relation to each other across the surface of a sphere such as our planet, their direction and rate of movement follow quite simple mathematical rules. There are on the surface of the sphere two 'poles of spreading', and an 'axis of spreading' passes through them. These are quite separate from the axis of rotation and its poles – which we call the North Pole and the South Pole. If imaginary lines of latitude and longitude are drawn in relation to the poles of spreading, plates always move so that transform faults run along the lines of latitude. Movement is slowest near to either pole and most rapid far away from them – that is, at the 'equator of spreading'.

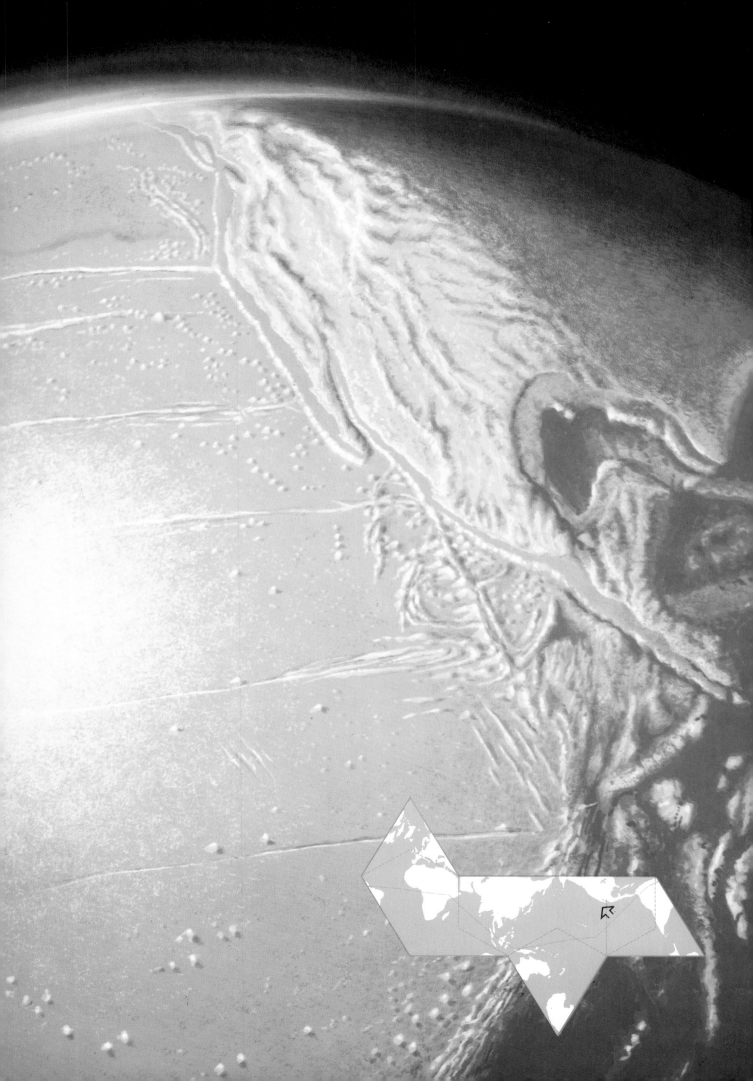

Abyssal plains

There are areas of the sea-bed deeps that are naturally dark, vast, flat and as desert and hostile as those words could possibly evoke: the lack of variety, the massive blanket of time passing uninterrupted while billions upon billions of microskeletons of radiolarians and plankton rain down. These are the abyssal plains, unbroken in the Atlantic and Indian Oceans for widths of 200 miles, though slightly relieved by hill provinces in the Pacific. As we cross back into the Atlantic region, we stop and look to the right, deep into the Guyana basin, to visit one of these.

The average gradient of these 'smoothest surfaces on Earth' is calculated to be 1 in 1,000, yet when soundings were taken they were found to be lying on bedrock that is far from even. What possible mechanism could have smoothed over such vast areas at depths that are not normally associated with currents and movement?

A deep grab finds the abyssal plains made up of sediments, but if these had gathered slowly by normal sedimentation the contours of the bedrock would have remained, even if softened in profile. The sediments in fact are found to be graded – that is, with layers of large to small particles alternating – indicating that they were laid at different times with enough energy to sort out the layers. The source of this energy seems most likely to be turbidity currents, a solution that came from observations first made in fresh-water lakes.

If fine sediments, such as clay, become suspended in water, they slightly increase its density. Even this small density increase will cause the water to sink, and as it does so potential energy (due to increased density) is converted into kinetic energy (the energy of movement) – hence the farther the sinking the greater the increase in speed. If the turbid water is a large volume, the friction will be small, so that the slightest increase in density brought about by the suspended mud or silt grains can lead to surprisingly high velocities – of the order of tens of miles per hour.

In 1929, an earthquake caused a turbidity current to tear through a number of deep-sea cables off the Newfoundland coast. This was proved in 1952 by Bruce Heezen and Maurice Ewing, who calculated, by timing the precise breaking points of the cables, that the sediments were being propelled at a rate of 45 miles per hour. Are such unpredictable occurrences taken into account before the dumping of chemical and atomic waste or cylinders of nerve gas in the oceans?

right A turbidity current rushes down a continental slope, cutting deeply into the sediments and rock and reaching a speed of perhaps 45 miles an hour. These massive movements of silt-laden water can exert a massive force, and when they reach the ocean floor spread the silts out evenly to form the flat, almost featureless abyssal plains.

Cuba

Great Bahama Banks

Exuma Sound

Bermuda Rise

Bermuda Island

Muir Seamount

0

1,000

left A profile of the North Atlantic ocean bed from Cuba to the Grand Banks shelf off the coast of Newfoundland passes through Bermuda. The vertical scale is exaggerated, so that Bermuda resembles a Gothic cathedral standing on a medieval plain, but this highlights the lonely, unique position of Bermuda. The island stands on a pyramid of rock and acts like a pivot for the ocean currents circulating round the western North Atlantic.

Grand Banks Shelf

0

20,000 feet

2,000 miles

Oceanic rises

Oceanic rises are the tenth and last of the typical ocean-bottom features to be encountered on our great-circle itinerary. Thrusting 14,000 feet up from the floor of the huge North American Basin is the volcanic pedestal of Bermuda. Its most striking feature is its central position. Like the hub of a huge wheel, this oceanic rise rears up from the ocean floor to surface more than 600 miles from the closest continental shore.

The oceanic rises are single elevated areas – rather than elongated ridges – rising at least several hundred feet from the sea bed. They have no apparent connection with either the continents or the mid-ocean ridge system, but rise up like isolated boils from the Earth's crust. There are three such major rises in the Atlantic: Bermuda and the Corner Rise north of the Equator, and the Rio Grande Rise to the south. Of these, only the Bermuda Rise was formed with enough strength and volume to mount its little islands above the waves. The Indian and Pacific Oceans have many such rises, some of them large, but many are yet to be named.

Partly due to its position, more is known about the Bermuda Rise than about most of the others. There are no standards by which to compare the persistent determination and force implied by the dimensions of this giant pedestal: it is calculated to contain 2,500 cubic miles of volcanic rock. The pedestal itself reaches within 120 feet of sea level, and the islands which break the surface are the result of millions of years of coral-production by tiny living creatures, together with ample deposits of calciferous sands swept up by Ice Age winds. The slopes of the pedestal fall away with the same starkness as the directional forces that moulded it, at gradients of 1 in 5 to 1 in 30.

When such massive weights of rock build up on an ocean bed, they must in time literally weigh it down. Where this has occurred, there are now flat-topped rises which are well below the level of the sea's surface. Such phenomena are called guyots, after their discoverer, Arnold Guyot. Some guyots in the Pacific have sunk by more than 3,000 feet.

The analogy between Bermuda and a wheel-hub is more than a fanciful image. The ocean currents of the North Atlantic – both the warm, shallow waters of the Gulf Stream and the deep, cold waters originating from the edges of Greenland – do indeed rotate about Bermuda as about a spindle. Following the Gulf Stream on its beautiful serpentine course – behaving remarkably like the meander of a giant river crossing a continent – our itinerary returns to its starting point to view one final oceanic feature.

Oozes, sediments and red clay

Before oceanographer Maurice Ewing devised the first deep-ocean echo-sounding techniques, only guesses could be made as to the thickness of the ocean-floor sediments. Therefore we can still share some of the shock experienced by some when Hans Petterson, leader of the historic Swedish Deep Sea Expedition of 1947–48, announced the discovery of Atlantic sediment layers 12,000 feet thick. But this sense of shock was due only to unfamiliarity with the timescales of planetary change. A natural comparison would be to think of a mountain with a comparable height. When we do take, say, Mount Pelmo, rising to a height of 10,000 feet in the Italian dolomites, not only can we compare the dimension but we can experience a second shock if unprepared: This mountain is itself totally built of ancient deepsea sediments, which must have been very much thicker before consolidation into rock.

Petterson and the Swedish oceanographers found in both the Indian and Pacific Oceans sediments no deeper than 1,000 feet. The reason for this is still uncertain; it may be linked with the rate of expansion of the ocean bed or due to quite hidden reasons yet to emerge. Professor Ewing has demonstrated how apt the snowfall analogy is in the drifting of the sediments among the foothills and peaks of the mid-ocean ridge of the Atlantic. The crowning rift valley, as expected considering its relative youth, is almost clear of sediments.

The types of sediments, oozes and clays that are found in various parts of the sea bed express their differing conditions, biological movements and geological behaviour. The broadcast division is between the near-shore sediments called *neritic* and the deep-ocean sediments known as *pelagic*. Those near to the great silt transporters, the rivers, naturally express the rocks, pebbles, boulders, gravel, silt and sand of the continuous breaking down of the landscape by wind and water action. Around the coasts one also finds many variously-coloured muds – white, black, blue, green or red according to their constitution, and apparently according to climate.

In the deep ocean are found massive oozes of living origin. Certain minute skeletal creatures have been shown to produce so much of the white sedimentary 'snowfall'; each is associated with different but overlapping regions. The foraminifers, their tiny shells perforated in a beautifully decorated manner, are a major group. The *Globigerina*, a genus of the foraminifers, divide in order to reproduce. Each time they do, they shed their shell and form two new ones – so they are producing sediment continually, not

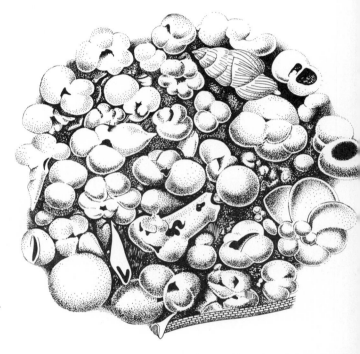

Except in the deepest waters, the thick oozes of the ocean beds are formed largely from the remains of microscopic animals and plants, and the three main types are shown here. **above** The delicately beautiful shells of radiolarians are made of silica, and are stronger than they look. Radiolarian ooze is found only in some tropical parts of the Pacific and Indian oceans, at depths of 6,000 to 30,000 feet. **above right** The pierced, chalky shells of foraminifers, such as *Globigerina*, form the most abundant oozes. But they are not found below about 20,000 feet because they dissolve under great water pressure. Both radiolarians and foraminifers are animals. **right** Diatoms, on the other hand, are plants. They have delicate silica shells, and are most common around Antarctica and in the far northern Pacific.

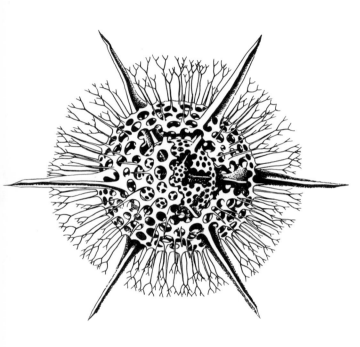

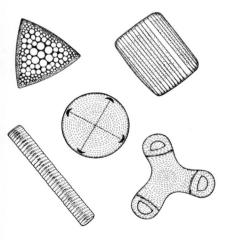

waiting to die to discard their skeletons. These last shells cover millions of square miles of ocean bed thousands of feet deep, so prodigious is their activity. They are mostly found on temperate sea beds, since in very deep water the lime basis of these skeletons becomes dissolved by the increased carbon dioxide content of the deeper water; in this case the minerals are recycled into the system before settling into solid bottom.

The minute diatoms, with their beautiful variety of semi-flat radially-symmetrical structures, flourish in colder waters. Hence there is a huge ring of diatomaceous ooze surrounding the continent of Antarctica. Another huge band stretches from Japan in a great-circle arc to Alaska. These minute plants contain silica, and become part of the nutrient-rich upwellings associated with the abundant sea life of both these northern and southern cold-water regions. The most beautifully clad of all these minute sea creatures are the radiolarians. They have been called three-dimensional snowflakes. Their structures exhibit the most sophisticated of all naturally occurring geometric forms. When the naturalist Ernst Haeckel in the last century drew his finds as he peered through his microscope, the established scientific world cast doubt on his integrity by implying that only he could see them.

Finally comes the mysterious red mud that covers more than half the solid surface of our planet. The minute particles that make up this deepest of sediments are only one-thousandth of a millimetre in diameter. The redness is due to the iron content. This is partly a result of land sedimentation but, in fact, the red mud has a higher content of both iron and manganese than continental rocks. Where does it come from? This mystery was solved when it was calculated that the planet receives between eight and ten tons of meteoritic iron, nickle and silicon from space every year! These minerals distributed in tiny pellet form and as a hydrated aluminium silicate, make the characteristic red muds of the deep ocean. The metallic content is also responsible for the nodules of both iron and manganese – varying in size from a pea to a potato – which are becoming valuable oceanic metal resources for man.

The sediments, which are in a continuous process of change, have been aptly compared to the leaves of a book that life and oceanic function is constantly writing on the sea bed. We have had our first peeps through the leaves of this book by drilling fine cores through the pages. We can expect in the future to learn a vast amount about the details of the past climatic conditions and the development of life as far back as the book will take us.

Chapter 4

Cycles of Change

A salmon leaps up the rapids of a river to its inland spawning-grounds. This annual pilgrimage of the salmon symbolizes the cyclicity of life, in which the oceans play a crucial role. Man's interference with the natural cycles through his industrial activities threatens the interlocking web of life itself.

A FLAME lasts only as long as the fuel it consumes. Life, ever since its first spark was struck by a complex set of conditions in the primaeval Earth, has depended for its survival and development on a constant renewal of its physical resources. Young and elementary life relied on a continual supply of the right proportions of chemical nutrients, warmth from sunlight and agitation from the movement of the waters. The rhythms of the great planetary circulations supplied these – and continue to provide the necessary conditions for life to flourish.

The great difference between life and flame, however, lies in the ability of life to influence its own destiny: to make its own fuel, to adapt to changing conditions and to affect the environment to its own advantage. The production of today's oxygen-rich atmosphere is believed to have been one of these effects. Although we may look upon life as an inevitable outcome of the prevailing cosmic conditions those misty aeons ago, it is important to realize the significance to the planet of such an event. Some scientists believe that organic action not only completely replaced the older atmosphere with the present one, but even built up the continental land masses by accretion and ancient oceanic sedimentation. The land, the air and the green fields man walks on may all have been laid down by the processes of life.

Life is not simply an addition to the face of the planet; it is a vital – *the* vital – link in the whole planetary system. Nowhere is this clearer than in the cyclic changes constantly renewing every vital feature of the physical environment. The cycling for re-use of water, carbon and nitrogen compounds and oxygen – the most important among the many expendable necessities of life – all depends on the co-operative interaction of the physical and living worlds. And playing a central role in this dynamic drama of a living planet are the oceans – truly the lifestream of the organism Earth.

The hydrocycle: water in circulation

Greatest of all the great cycles of change is the hydrocycle, the worldwide circulation of water. Water shows itself on Earth in a great variety of forms – as sea, vapour, snow, ice, rain, stream, lake, river-estuary, ground-water. All, however, are stages in a constantly renewed circulation, although the stay in a lake-bottom or glacier is much longer than in a fast-flowing river.

The motive power of the hydrocycle is the heat of the Sun. Sunlight both evaporates masses of water from the seas to create cloud and also powers the process of trans-

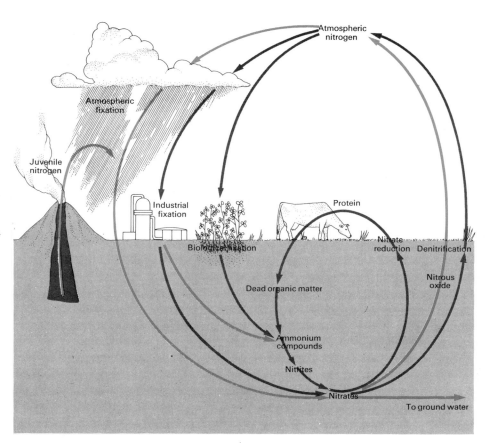

right The nitrogen cycle: Almost 80 per cent of the atmosphere is nitrogen, a chemical element vital to life processes. But atmospheric 'free' nitrogen cannot be used by living things; it must be 'fixed' – converted into substances such as nitrates, which plants can convert into proteins. Fixation can occur in the root nodules of legumes, such as beans, in the atmosphere in thunderstorms, and by man's synthetic production of nitrogen-containing chemicals, so extensively misapplied as fertilizers. The bacterial breakdown of dead plants and animals and their wastes also produces nitrogen compounds, but other processes are also constantly returning free nitrogen to the air. The process is a ceaseless cycle involving land, sea and atmosphere.

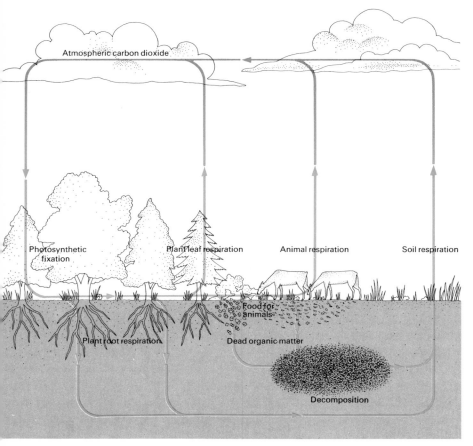

left The carbon cycle: The use and re-use of carbon – the fundamental chemical element of life – is also cyclic, but the carbon cycle is simpler than the nitrogen cycle. There are two main processes: photosynthesis, by which green plants 'fix'. carbon by converting atmospheric carbon dioxide into food materials, and respiration, by which both plants and animals break down carbon compounds and release carbon dioxide into the air. Bacterial decomposition is a special form of respiration. Also important, but not shown here, is man's burning of fossil fuels, such as coal, which also releases carbon dioxide into the air. Many scientists believe that sea life plays a much greater part in the carbon cycle than life on land.

piration by which large amounts of water are drawn from
the soil by plants, to be given up as vapour from their
leaves. There are, correspondingly, two types of cloud-
bearing air mass. The more heavily-laden maritime air
mass brings not only the continental water supply in the
form of rain, hail and snow, but also a good deal of latent
heat, representing the energy needed to evaporate the
water. The continental air mass has lighter cloud made up
from a number of sources, including evaporation from
falling rain, damp soil, ponds, lakes and rivers, and
transpiration from vegetation.

A most important aspect of the hydrocycle is the way
in which the distribution of rain performs several simultaneous functions. It supplies the land flora and fauna with
their vital water, but it also assists in breaking up the stark
rocks, forcing them to give up their valuable raw chemicals
to feed life processes. This erosion is accomplished by
the driving force of the raindrops breaking up the rock
face, and also by the action of water that percolates into
cracks in the rock formed by temperature changes: when
freezing weather comes, the expansion of the water bursts
open even the most powerfully bonded rock. The results of
this phenomenon can be seen wherever a sheer-faced
cliff-top has become grass-bearing topsoil.

Having fallen as rain, the solvent power of water brings
nutrients won from the rocks to the lower lying soils and
vegetation. The run-off becomes a stream. The saturated
soil and porous rock hold reservoirs of ground-water, which
feeds artesian wells and emerges in hillside springs when it
reaches an impervious layer of subsoil or rock. The streams
supply animal life, including man, with drinking water.
They also begin to pick up silts and deposited wastes and
carry these to the rivers and on to the sea. Depending on
the density of human habitation that they pass through, they
may become so overladen with waste materials, organic and
inorganic, that they become a source of poison to the seas
they empty into rather than the life-bearing mineral waters
they originally were. This polluting aspect of the hydrocycle will be examined later in this chapter.

Estuarine waters are biologically the richest waters of the
planet in normal circumstances. The subtleties of the mixing
of sweet and saline waters in the protected shallows of an
estuary have been the subject of much scientific and
artistic attention. The rivers bring to the sea invaluable
nutrients, and the estuary shallows are the meeting place
of sea and land life. The movements of the shallow waters
are of great benefit to the smaller planktonic (floating,
drifting) and benthic (bottom-living) green life which, by

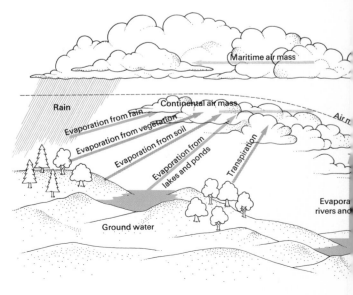

ration from sea

Sea

above Turner's painting *Ulysses Deriding Polyphemus*, apart from its mythological content, demonstrates in great beauty the main elements of the hydrocycle: sea, Sun, clouds, rain and land. The story of Ulysses symbolizes the archetypal psychological journey through life, and it has been suggested that Turner has echoed this in the language of landscape as the journey of the life-nourishing waters. **left** The drawing shows how water is separated from the main body of the oceans by the Sun's power, to be distributed and released as individual droplets onto land. Eventually, via streams and rivers, it returns to the great oceans, just as Ulysses returned to his homeland.

making food through photosynthesis, creates the basic diet of the rest of the marine food chain. These protected continental waters also provide spawning grounds for large carnivores – larger fish which eat smaller sea life like shrimps – because of the abundant planktonic food supply, plant and animal.

The hydrocycle completes its never-ending chain as the estuary waters slowly diffuse and join the major ocean currents to recirculate around these planetary arteries. It is only a very small proportion of the total ocean that is being taken to the land and back – a mere 1.5 per cent of the whole global water supply. And this is calculated to be actually airborne for at most only twelve days. Yet its contribution to the web of life is incalculable.

The tides' ebb and flow

We have seen how the Sun's energy in the form of heat pulls the ocean waters into the atmosphere by evaporation. Now the Moon joins in the pulling game, and we get both solar and lunar tidal effects on the Earth's surface. Their pulls are exerted everywhere, but since the waters of the globe are most sensitive to their effects, they show them most dramatically.

In order to understand fully the basis of the sea's tidal rhythms, it is imperative to remember the importance of planetary rhythms that are normally taken for granted – the daily rotation of the Earth on its axis, its yearly rotation around the Sun, and the Moon's rotation around the Earth. We generally measure these rhythmic periods of rotation in terms of their appearance to us, standing on the Earth's surface, which is itself in motion. As a result, the apparent periods are different from the true periods measured in relation to the 'fixed' stars. So, for example, the daily rotation of the Earth takes 24 hours by our own Sun-measurements, but 23 hours 56 minutes 4.09 seconds by precise star-time. In the same way, the Moon actually orbits the Earth once every 27.3 days, but the period from one new moon to the next, through the phases of waxing and waning, is 29.5 days. The difference is due to the Earth's own orbital movement through space. And the time from one moonrise to the next – the lunar 'day' – is 24 hours 50 minutes.

The Moon is a mere speck of cosmic dust in comparison with the Sun, but it is some 400 times closer to us. Gravitational attraction increases in direct proportion to mass, but falls away in proportion to the square of distance (that is, distance multiplied by itself). So it is the Moon that wins the cosmic tug-of-war, and imposes its rhythm on our

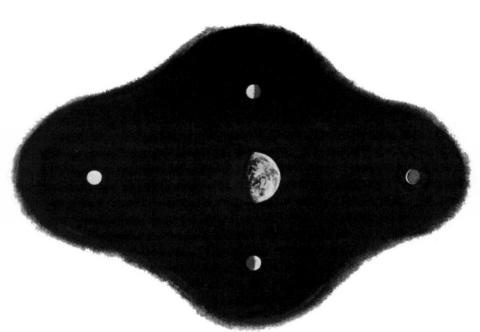

left The tides are caused by the combined gravitational forces of the Sun and Moon. The Earth is shown here illuminated from the right by the Sun, but the Moon is shown in its four phases as seen from the Earth's surface. The tides are greatest when the Sun, Moon and Earth are in line and slackest when they form a right angle, as at the first and last quarters. **bottom** The map shows the worldwide pattern of tidal ebb and flow. The lines join points at which high tide occurs at the same times. The lines converge on points called tidal nodes, or amphidromic points. Here, there is no rise and fall of the waters, but they flow vigorously to and fro. This phenomenon is caused by the land masses interfering with the tides. The nodes would not exist if the world were totally covered by water. The darkest tinted areas on the map are those with the greatest potential for generating tidal power.

ocean waters: the tidal periods, although they vary according to local topography, correspond approximately to the lunar day.

The gravitational pull on the waters is greatest on the side of the planet nearest to the Moon, and least (by the inverse-square law) on the opposite side. Hence the waters bulge away from the Earth's centre on the opposite side, too, and high tide in most places on the planet occurs twice daily – every 12 hours 25 minutes. It can be shown that, taking into account mass and distance, the Sun has a tidal pull about half that of the Moon. But this half can be dramatically effective. When the two heavenly bodies reinforce each other – at new moon or full moon – we get the great spring tides. (They have nothing actually to do with the season of the year.) When the situation is reversed and the pulls of Sun and Moon are at right-angles to each other – at the first and last quarters of the Moon – the tides are at their slackest. These are the neap tides. Thus the tides follow faithfully the phases of the Moon.

But how do the tides affect man and other living things? Are they of concern only to mariners and fishermen, yachtsmen and bathers, and the inhabitants of sea-shore rock pools? Or does the Moon on its daily passage overhead influence more than just the oceanic part of the hydrocycle? The existence of rhythms in the living world has been known for a long time. There are old country traditions that

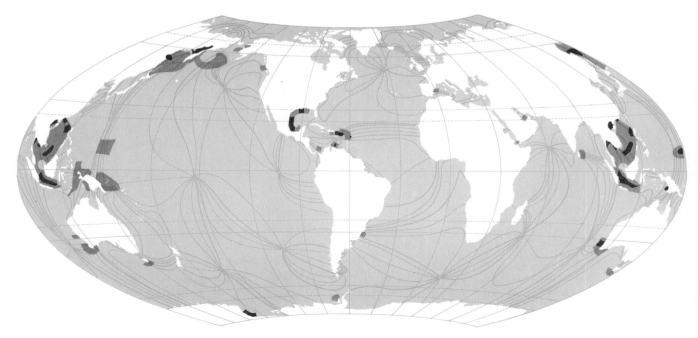

plant growth takes place at night, and even in 1660 a paper delivered to the Royal Society of London spoke of shellfish of the East Indies being 'plump and in season at the full moon and out of season in the new'. Contracts for the sale of mahogany in Cuba traditionally had a 'moon-clause' specifying when the wood should be cut to ensure minimum moisture, and hence maximum hardness, in the timber.

More recently, the biological world has been puzzled and excited by the discovery of circadian (daily) rhythms that are apparently independent of obvious outside clues as to the time of day – such as light and dark. As long ago as 1729 it was noticed that the daily rhythms of leaf movements in certain plants continue even in constant darkness. Now it is known that a multitude of bodily functions in man and other organisms display rhythmic variations – including sleepiness, hunger, pulse rate, body temperature, oxygen consumption, carbon dioxide production, urine excretion and others. Each cellular system, it seems, has its own physiological clock. The exact way in which this clock works and where it is located remain a mystery. Nor is it clear whether the timing mechanism is internal, or whether organisms respond to external changes. If the latter, could it be that the Moon is implicated?

Some remarkable facts have emerged from various laboratories. The Swedish scientist Svante Arrenhius has demonstrated how the growth of the human egg follows a period of 27.3 days – the orbital period of the Moon. And as the Moon passes daily overhead, there is evidence of its having a tidal effect on the liquid contents of cells of men, plants and animals. American physiologist F. A. Brown reported in 1954 the result of taking some oysters 1,000 miles inland from Long Island, New York, to his laboratory at Evanston, Illinois. These oysters kept up their valve-opening habits corresponding to the tides – or lunar days. But then, after a slowing of the rate, Brown was amazed to find that the oysters were opening their valves in time with a new tidal rhythm. This corresponded to what the tides would have been at Evanston if this had been a coastal town. The oysters were apparently responding to the daily passage of the Moon over the local meridian.

Other scientists interpret the observations in a different way, among them Professor Erwin Bünning of Tübingen University. Noting that unicellular organisms as well as the cells of higher plants and organisms exhibit 24-hour rhythms and that these rhythms continue in laboratory conditions without tidal changes or moonlight, he states that this 'strongly suggests the existence of an endogenous [inbuilt] lunar rhythm'. Here are two scientists in different laboratories studying the same kind of phenomena. Both link the rhythms of the Moon – which produce the ocean tides – to the rhythms of living organisms. One proposes an inbuilt mechanism, while the other demonstrates behaviours that suggest a response to external, environmental factors. In Brown's own words, 'The various clocks within the organism are tuned by the common rhythmic geophysical environment of this planet.'

Could we speculate that the inbuilt, instinctive timing mechanisms that man has inherited from his predecessors might constitute an evolutionary parallel to the way in which the ocean floors carry the imprint of the Earth's magnetic field at the time they were born? Could it be that the 22-hour rhythm of algae, the 23.3-hour rhythm of oat plants, the 23- to 26-hour rhythms of animals, and the 27-hour rhythm of bean-leaves are also the imprint of the lunar tidal cycle at the time in evolution when they each emerged? As our bloodstream apparently echoes the constitution of sea water, did we also adopt or inherit the tidal rhythms for each cell in our body?

The Sun's contribution: energy
If the oceans are the planetary lifestream, what does the whole Earth-organism live on? What is its basic food? Almost the only source of nourishment is energy streaming from the Sun, a small fraction of which the Earth borrows before re-radiating it to space. We ought therefore to consider the Sun's role a little further.

The Sun not only induces the planet to follow it around the Milky Way, but it also supplies the motive force for all the Earth's multitudinous activities. The ever-flowing flood of various wavelengths we receive as sunlight are almost our only source of life. But raw solar radiation is too fierce to be utilized undigested, and the planet has on the outer fringes of its enveloping atmosphere its own digestive system. This has two parts. In the early formative years of the planet, it is believed, the molten metals of the core set up extensive magnetic fields through a dynamo-like action. These huge fields protect us from the intense bombardment of charged atomic particles streaming from the Sun – the 'solar wind'. Later in our planetary evolution, the changing atmosphere developed a high-altitude screen of ozone molecules. These absorb most of the biologically harmful ultra-violet wavelengths from the flood of radiation.

Thus the range of wavelengths that principally affect the Earth extends from the ultra-violet through the visible rays to the infra-red. This last is radiant heat, which warms both land and sea. The land responds with rapid temperature fluctuations, but the sea has powerful heat-retaining properties. Both the land and the sea heat the air masses by conduction and convection, and by return radiation. This heating of the air, together with the differing rate at which the water surface is heated due to the Earth's curvature, cause movement from the equatorial regions towards the poles. The movements in the waters directly influence the movements of the air masses above, and the evaporation of sea water by solar heat contributes to an increase in water density, which also causes flow towards the poles. The spin of the planet on its axis also comes into play, affecting the behaviour and direction of the water masses as they flow towards the temperate zones. This is the coriolis effect. The whole organism is set into circulation and life by the donation of solar energy, part counterbalanced by the proximity of the Moon and part moved by the natural spin of the Earth.

Movements in wind and water

How powerful an effect the weather has on each of us! The working day of the farmer, builder or postman may be totally made or marred by the behaviour of the sky. Families who have looked forward to a sunny holiday for a whole year can be bitterly disappointed if the isobars fall unpredictably low and bring rain for a week at a time. In the same way, winter sports and outings can be blessed by unusually fine, sunny weather. It has been said that every

man brings to work with him the morning's weather. But how many of us place the blame for the weather where it really belongs – in the oceans? It is the wavelengths of radiant heat that reach the Earth from the Sun, and the capacity of the oceans to absorb them, that start into motion the huge atmospheric engine. An apt morning's greeting might be, 'The Atlantic has been kind today.'

But the movements of air and water masses play a far greater role in the story of life than merely by influencing the weather. Animal life in the sea, as elsewhere, relies completely on the presence of oxygen. In the oceans this is in a dissolved form. Sea creatures consume oxygen which must be replaced by the oxygen-producing phytoplankton (planktonic plants) and other marine plants, or with oxygen absorbed from the air. There is a limit, however, to the quantity of oxygen that can be dissolved in sea water; surface waters can reach a saturation level. If this layered water were not circulated by currents, the deep waters would have no vital oxygen. The upper layers, on the other hand, would lack the equally vital mineral upwelling from the sea bed. The existence of life throughout the ocean depths shows that thorough mixing does occur.

The mixing of water masses goes hand in hand with the mixing of air masses. The circulation of oxygen is assisted by the presence of the warmer water due to insolation (the warming of the sea by the Sun). The heat taken up into the atmosphere in vapour is only released when the

left The power source that maintains life on our planet is the Sun. Its radiation creates all our food supplies by powering the process of photosynthesis, and the long-wavelength infra-red waves bring us warmth. The Sun's bounty is filtered, however, by the ionosphere, the upper layer of the atmosphere where many gas molecules are converted into ions, and by the lower stratosphere, where oxygen is found as ozone. These layers absorb the strong ultra-violet rays, X-rays and gamma-rays poured out by the Sun. Without their protection, all life on the surface would be fried.

evaporated water condenses in the air to form cloud and rain. The sea is also constantly losing heat by long-wavelength infra-red radiation, much of which is trapped by water vapour and carbon dioxide in the atmosphere. This delayed-action heat-loss helps keep the night side of the planet warm. Water vapour is a lighter gas than both oxygen and nitrogen – the main components of the atmosphere – and thus it tends to rise upwards. Since the atmosphere is heated from below by land and sea, its temperature drops with altitude until the ozone layer is reached, where strong ionization due to solar radiation causes a rise in temperature once again. Thus as water vapour rises and encounters the cooler altitudes it condenses into cloud.

The Earth's gravity affects both sea and air, resulting in an increase in pressure as one gets nearer to the Earth's centre. Pressure gradients in the sea and air are remarkably parallel. Sea and air distribute themselves in currents which also have similarities. The main difference is that the land masses interfere with the water flow much more than with the air flow – although temperature differences over land do markedly affect the winds. Both wind and water masses move away from high pressure and into low pressure.

The overall planetary pattern of the atmosphere takes the form of a series of regions or cells. A cross-section through the whole system shows twelve such cells, which rotate in different directions according to whether they are in the northern or southern hemisphere. The spin is again

the product of the Earth's rotation on its axis, manifesting itself in the coriolis effect. Low-pressure areas – cyclones – in the northern hemisphere cause an anticlockwise spiral circulation inwards, while high-pressure areas, or anti-cyclones, cause the air mass to spiral out from the centre in a clockwise direction. The circulations are reversed south of the Equator.

A beautifully complex relationship exists between wind and waves, caused by a combination of friction, pressure gradients and coriolis force. The resultant direction of the wind in relation to the water is called the geostrophic wind. Wind speed is greater the closer together the pressure differences are. Greatest weather changes take place where the pressures – high or low – fall off most rapidly. This occurs where the warm air of the temperate zones meets the more dense cold-air regions – hence the Roaring Forties of the southern hemisphere and the westerly gales of the northern. At the Equator, with its unchangingly warm air, are the dreaded windless doldrums, which broke both the heart and the purse of many a sea captain in the days of sail power. But just as there are the strong currents that channel their way through the oceans – the ocean arteries, as Matthew Maury called them – so there are similar phenomena in the upper atmosphere. These are the jet streams, which reach speeds of more than 200 miles per hour. They represent the most extreme examples of mixing and circulation in the worldwide air and water cycles.

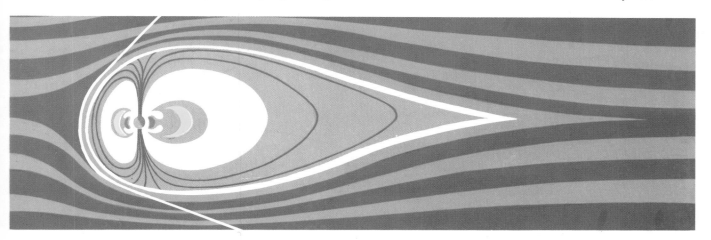

above The shape of the magnetic sheath surrounding the Earth has been revealed by scientific satellites; this is the space-age image of the organism Earth, showing that the oceans and all life-forms are well within the total planetary body. The magnetic field and the solar wind – the stream of electrically charged particles streaming from the Sun – interact to distort the basic doughnut shape by giving it a long tail on the night side of the planet. But the sheath serves a vital filtering function: it traps the charged particles that would otherwise mercilessly bombard the Earth's surface.

Currents in wind and water are inter-
related and are the result of many forces,
including the Earth's rotation, heating and
cooling as a result of solar radiation, and the
Coriolis force, which deflects north- or
south-flowing currents to one side. The
Earth spinning on its axis creates cells of
circulation. **top** Because of the continents,
these cells in the oceans turn back on
themselves. **bottom** The air, however, has
no impediment to its flow, so that the cells
can circulate right around the globe. The
only place where this can happen in the
oceans is in the Antarctic, where there are no
continental barriers. **centre** In the Atlantic
and Pacific (and to a lesser extent in the
Indian Ocean), the surface currents form
roughly symmetrical patterns. Circulation is
clockwise in the northern hemisphere,
anticlockwise in the southern.

Weather watching

Predicting the weather has been a natural human preoccupation for as long as man has been aware that certain heavenly clues indicate that certain climatic events are imminent. Forecasting has developed from intuitions through rules of thumb to our modern computer analyses of atmospheric data. In spite of the sophistication of these last, there remain just three main types of sky conditions. If high air pressures are prevalent, we generally have clear skies and sunshine. If low pressures are the order of the day we get a mixed bag of wind squalls, rain and possibly storms. The other major alternative occurs if we are on the border of the pressure gradients, when a cloud division may cross the sky from horizon to horizon.

But how are clouds formed? Sun-heated sea causes vaporization, to form the gaseous state of water. The atmosphere, however, can take up only so much vapour. This maximum is called the absolute humidity. It depends entirely on temperature: the higher the air temperature, the more water vapour the air can hold. When saturation point is reached the vapour begins to condense into cloud and further into water droplets. This commonly occurs when water vapour (which is lighter than air) rises. On a simple ocean/land profile we can see cool, moist air blowing inland. Warm air from the heated land surface takes this sea-breeze up to cooler regions where clouds form. This is what happens in day-time. At night an opposite flow of air results from rapid land cooling bringing a weak land-breeze out to sea, and the warmer sea causes a slight uplift.

There are four main reasons for atmospheric cooling and condensation. First, radiant heat lost from the land at night causes a fall in ground temperatures and a corresponding fall in atmospheric temperatures. This is related to the formation of dew by the saturated air, and is reduced if cloud cover reflects back the radiant heat. (It is for this reason that clear nights are cool.) Second, the rise of thermals (ascending parcels of warm air) results in the expansion of these air masses due to the lower pressure at higher altitudes. This expansion uses up the internal heat-energy of the air, resulting in cooling. Third, the ascent of airstreams over topographic barriers – cliffs, hills, mountains – and over wedges of cold air nearer the ground results in a similar expansion and cooling. Fourth, air masses at different temperatures may meet and mix, as at cold and warm fronts, causing cloud formation.

According to the atmospheric conditions, clouds take on a wide variety of shapes. Their names are taken from Latin descriptive terms, among them *stratus* (layer), *cumulus* (heap), *cirrus* (fibre), *nimbus* (umbrella), *altus* (high), and *fractus* (broken). The shape and form betray their ice or water content and the conditions inside them, and these in turn give important clues to weather conditions. The formation of cloud layers in the air is analogous in transportation, diffusion and turbulence to the patterns found in the ocean waters. Both atmosphere and oceans show such features as layering, vertical and horizontal mixing, and the formation of eddies.

Currents and counter-currents

The great internal convection currents of the Earth's mantle, which created the ocean floor, well up at the mid-oceanic ridges and spread laterally until they are pulled back into molten mantle via the deep trenches. This is a multimillion-year cycle quite beyond the appreciation of man's senses. Like the bones in the human body, the crust grows slowly and steadily in comparison with the throbbing, racing circulations of the fluid systems. Similarly, the ocean circulations are a rushing, gushing affair in relation to its solid boundaries. But with all types of circulation, for one part to move in any direction means that another equally large part must give way to allow it. So we have to think in at least two levels – of movement and counter-movement – in the oceanic circulatory system. These divide basically into warmer top currents and colder deep currents.

Knowing a general ocean-current pattern is like knowing a general climate, but day-by-day variations are more like the weather – extremely variable and almost defying prediction. The Gulf Stream, which follows a beautiful general sweep across the Atlantic from the West Indies to northern Scandinavia, when seen in daily detail writhes around like a gigantic sea monster with many heads. These long filaments, when charted intensively, were seen to be narrower and more classically meandering than any oceanographers had imagined. This meandering, with its huge eddy patterns, can over a period of ten days sweep through motions that isolate an island of some ten million tons of subarctic water and inject it with all its life into the subtropical Atlantic.

Left to itself, without continental masses, the worldwide ocean would probably circulate right around the globe along lines of latitude. Where it can, the ocean currents do just this – around the Antarctic, responding to the prevailing Roaring Forties winds. However, there are massive land barriers which cause the major currents of the surface to spin in a predominantly westerly direction. The spherical

A cyclone photographed from space by the astronauts of Apollo 11 demonstrates in a majestic way the three-dimensional circulation of weather forces. The air and the water vapour it carries are in constant flux from minute to minute, gyrating between areas of high and low atmospheric pressure.

shape of the Earth sets up a natural diversion for this westerly current by directing it to the right and north in the northern hemisphere, and to the left and south in the southern hemisphere. The exceptions to this rule, assisted by the coriolis effect, are the equatorial counter-currents of the Indian and Pacific Oceans.

To someone investigating the global distribution of the production of primary sea life, it is immediately evident that these great east-flowing currents are directly responsible for the presence of huge life-pools where they meet the western walls of their respective continental masses. The simple explanation is that the currents take up the nutrients when they meet the continental slopes off the coasts, and set up ideal fertile waters for all members of the marine food chain. These waters are associated with the Canary Current off north-west Africa, the Benguela Current off south-west Africa, the Peru Current off the South American Pacific coast, the California and Alaska currents off the west coast of North America, and the Gulf Stream with its Irminger and Norwegian branches off the West European continental shelf. The other huge production area for sea life is the West Wind Drift current that circulates all the way round Antarctica.

The web of ocean life
Life in the sea means food for other creatures – man included. And when we consider oceanic food production rather than the production of life or the natural balance of life, it becomes more and more apparent that we must regard man as part of the oceanic food chains. Such an approach will be viable, however, only when we learn what we have to put back into a system when we take something out – particularly where that system is as subtle and time-tried as the food chains of the ocean waters. It is inevitable that the laws of mariculture will have to be as rigorous as a scientific discipline if we want the waters to remain viridian with life and not the desert blue of a dead ocean.

What are the principles of ocean fertility? They do not differ fundamentally from those applying on land. There are generally five levels of energy consumers, each supporting the other in a chain and all ultimately supplied with motive power from the Sun. The primary or first level comprises the photosynthesizers. These are the phytoplankton, the microscopic plants that produce organic material – food – from sunlight, air and minerals by photosynthesis. They are food for the next level, the herbivorous zooplankton, such as minute copepods and crustaceans. These in turn are food for the third level, larger carnivores such as crustaceans and fish, which are often followed by another level of carnivores or, as on land, of omnivores which may feed right down to the bottom of the chain. (This fourth level is more common on land; bears and man are good examples.) The final level in the chain is often forgotten. It comprises the bacteria – largely bottom-living – whose job it is to break up dead organic matter, the falling dead bodies of the larger members of the food chain. This fifth level completes the cycle, releasing mineral nutrients which in turn await water movements to bring them up to the surface waters for the phytoplankton to feed on.

These, then, are the members of the food chain; what are the conditions necessary for living things to flourish? Not forgetting the basic requirement that the environment must be seeded with life in the first place, these fall into three basic groups: the physical conditions of heat, light and motion; nutrients; and biological climate – that is, biologically supporting associations between creatures.

The physical conditions are partly related to position. The amount of light a creature receives and the temperature of its surroundings are related to its depth below the surface and position in relation to moving water masses. Salinity, or salt concentration, is also closely related to this. Movement is linked closely with topography. Shore contours and island profiles assist greatly in producing circulation, upwelling and valuable mixing. River run-off,

wave mixing, convection circulation caused by the Sun's heat, eddies and turbulence, and other factors all contribute to water motion. And motion in turn aids in the distribution of nutrients. The main ones necessary, apart from the water and carbon dioxide used in photosynthesis, are phosphates, nitrates, silicates and traces of iron, manganese and some organic accessory growth factors. These are then synthesized by sunlight into organic compounds. The final requirement – biological cooperation – strikes at the core of the problem of man's relationship with the oceans, and must be examined in greater detail.

Completing the cycle
To the scientist, the objective dealer in data, it is quite clear that the fish that are taken out by man can seriously upset the energy budget or life-balance of an oceanic area. But to a commercial enterprise with competitive hunters on the seas there is little to restrain the overkill. For the politicians with huge home consumption markets, often living below protein subsistence standards, the incentive to implement controls is almost nil. Yet somewhere along the line a cost-benefit analysis has to be made that includes a realistic time-scale. Already the list of overkilled fish in the North Atlantic alone covers six species and twenty-four fishing grounds. It is estimated that, over the fishing grounds from Norway to Greenland to Nova Scotia, the cod, plaice, herring, hake, haddock and ocean perch are all

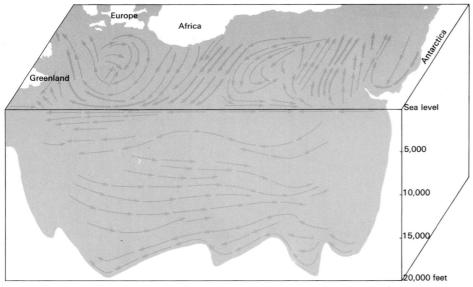

right This map of the North Atlantic water movements shows many more of the subtleties of ocean currents, but even on this scale relatively little detail can be shown. The Gulf Stream, for example, meanders so much that each arrow on this map should be replaced by many, showing a serpentine course with whorls and eddies that are constantly changing. Any map can show only the main trends. **left** This three-dimensional diagram of the Atlantic illustrates oceanographers' increasing knowledge of the intricacies of layered waters and their flow. Such knowledge is relevant both to an understanding of ocean life and to warfare, for submarines can use certain layered water effects to escape detection.

below This map of the main surface currents of the oceans shows that there are three major girations in the southern hemisphere and two in the northern hemisphere. The circulation in the Indian Ocean north of the Equator is much smaller than the others, and is also profoundly affected by the seasonal monsoon winds.

Key to principal currents : *1* North Equatorial Currents. *2* South Equatorial Currents. *3* Equatorial Counter-currents. *4* West Wind Drift. *5* Gulf Stream. *6* North Atlantic Drift. *7* Labrador Current. *8* Canaries Current. *9* Guinea Current. *10* Brazil Current. *11* Benguela Current. *12* Mozambique Current. *13* Monsoon Drift. *14* West Australian Current. *15* Kuro Shio. *16* North Pacific Drift. *17* California Current. *18* Alaska Current. *19* Oya Shio. *20* East Australian Current. *21* Peru (Humboldt) Current. *22* Falkland Current.

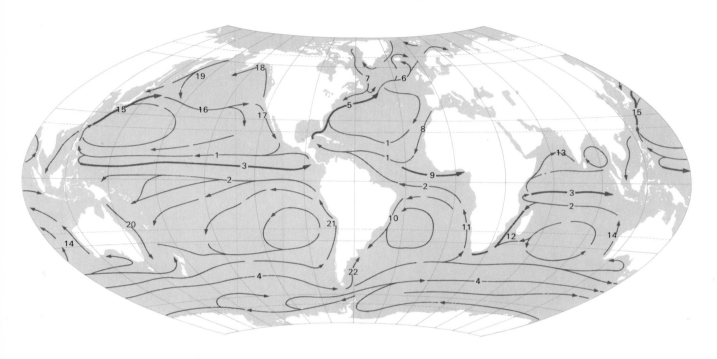

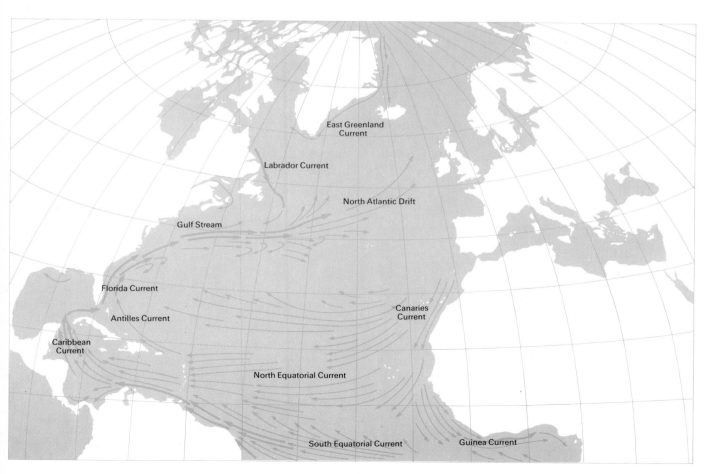

incapable of recovering naturally from the pressures of sustained hunting until international agreements are implemented to reduce catches.

But this is only one end of the problem. The other is the over-simple fact that the estuary waters of the coastal regions are the traditional spawning grounds of much basic sea life, and it is just these estuary waters that industrialized man has most heavily polluted. How can one species – man – expect a larger and larger slice of the limited energy budget of the food chain if he not only refrains from putting back into the chain due recompense for what he removes, but goes to the opposite extreme and feeds poisons into the system? American environmental campaigner Barry Commoner pleads for an enlightened industrial ecology by using the catch-phrase 'closing the circle', emphasizing the need to recycle and re-use the products of our industrial civilization. But this circle has really to be seen as a sphere, and a sphere has depth. This means that we have no escape clause for our discarded wastes – whether they are the poisonous by-products of industrial processes or simply products that have reached the end of their useful life. Either the pollutants themselves or our advertising-stimulated wants, whose satisfaction gives rise to the pollution, have to be neutralized before the products can be pronounced fit to be discarded to nourish the tissues of the Earth-organism.

One of man's most dangerous twentieth-century myths is the 'out of sight out of mind' policy that worked when all materials were biologically based or corrodible metals. The pandora's box of discoveries in high-polymer organic chemistry has led technology-intoxicated man into the realms of non-cyclic materials. The epithet 'everlasting' seemed so attractive until plastics made it almost come true. Now the unbreakable, incorrodible bottle floats in the most remote of wild seas carrying no other message than the foolishness of man. It is a monument to all the irreplaceable fossil fuel consumed to make a throwaway triviality whose useful life was a millionth of the time it took natural processes to create its raw material. It floats as a reminder that the constituents are locked in a non-cyclic limbo.

'To destroy the purity, make foul or filthy'

Thus a typical dictionary entry defines pollution. It provides a direct beginning to an examination of ocean pollutants, but one of the greatest difficulties of oceanic pollution is that the 'foulness' or 'filth' is far from obvious. The dilutions are so great that the pollutants are often invisible to our eyes, but are lethal to the microscopic life-

A fish is captured, paralysed and eaten by a sea anemone. All animal life continues to exist only by eating other life; hence the predator/prey relationship exists right through nature. But man, with his superior killing ability, threatens to take out of the system more than is replaced by natural processes, thus upsetting the whole delicate balance.

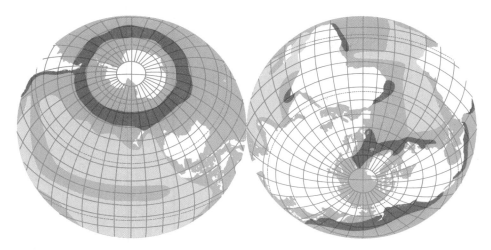

below A typical marine food chain headed by the adult herring illustrates the complex and many-staged feeding relationships that exist in the seas. The various creatures are not drawn to scale, and the food chain is not, of course, complete. The herring itself is the prey of many carnivores, including sea birds and the arch-predator, man.

above All marine life depends ultimately on phytoplankton – minute green plants that make food by photosynthesis – and this is illustrated by this map showing global food-production potential. The darkest areas represent the richest sea life due to the greatest phytoplankton growth, which is in turn related to the greatest movement of water bringing up chemical nutrients.

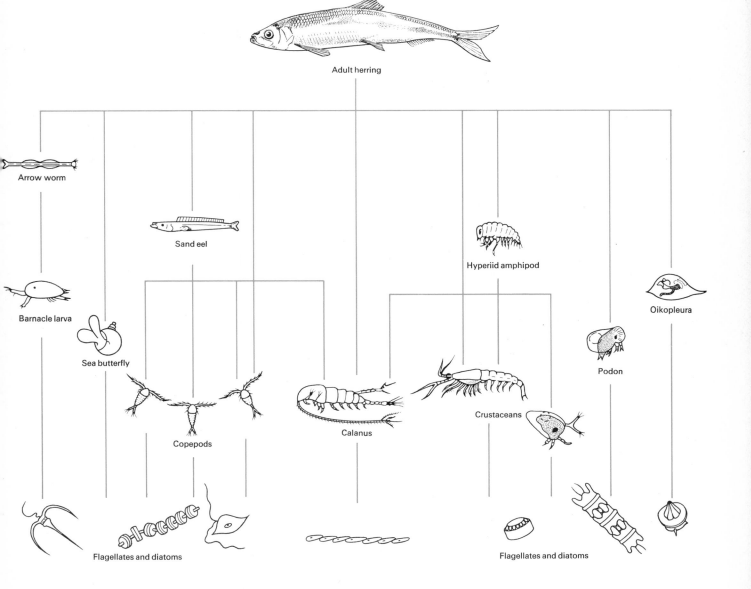

Adult herring

Arrow worm

Sand eel

Hyperiid amphipod

Oikopleura

Barnacle larva

Sea butterfly

Podon

Copepods

Calanus

Crustaceans

Flagellates and diatoms

Flagellates and diatoms

Oil-covered crew members of the tanker
London Valour cling to lifeboats after the
wrecking of their ship off the Italian coast
in May 1970. This was just one of many
such accidents occurring as growing
numbers of giant tankers began to cross the
oceans in the 1960s in response to man's
ever-increasing thirst for oil. Pollution from
such spectacular accidents is widely
publicized. But it is minor in comparison to
that resulting from the unpublicized regular,
deliberate dumping of crude oil as tankers
clean their tanks. Such pollution cannot help
but endanger the marine food chains as well
as foul the beaches.

forms that are the basis of the whole life-food chain. When things become really bad, then the foulness is not only nauseatingly obvious but it already represents a multi-billion-fold disruption of a natural cycle. As previously shown, dispersion of materials in water is fast and global. As a result, additions to the oceanic environment introduced by man threaten to be more lethal pollutants than those dumped on land. The following represent pollutants that are currently disrupting life:

Chlorinated hydrocarbons   These chemicals are best known in the form of DDT, which was so extensively applauded as a disinfester at the end of World War II. It has since graduated into a general pesticide, and has found its way into the seas with a vengeance. It is concentrated up the food chain, and is now found in the tissues of all living creatures. A total ban is impossible, it is claimed, because it would halt the fight against malaria. But this use is a far cry from the indiscriminate spreading of tons of DDT on farmland. Closely related and recently incriminated pollutants are polychlorinated biphenyls, or PCBs, which are widely used in industry.

Heavy metals   These include copper, zinc, chromium, cadmium, nickel, lead, mercury and iron. Many of them find their way into the sea via estuarine waters as the run-off from mine workings, waste tips, extraction and smelting plants, and other industrial processes. Car exhaust fumes have been incriminated as a major source of lead pollution that originates in anti-knock additives to fuel. The heavy metals are lethal to land and ocean life alike, but the threat to the phytoplankton is particularly grave, for these microscopic plants produce much atmospheric oxygen. Mercury pollution was seen to come home to roost on man's table in 1970, when some canned fish was found to contain dangerously high levels.

Petroleum products   Crude oil is now found in every part of every ocean, killing marine life more extensively than most people realize. Disasters of the *Torrey Canyon* type reach the headlines, but more widespread damage is done by deliberate dumping as giant tankers wash out their tanks. Underwater oil drilling poses another, growing problem, with the threat of well-head blowouts.

Municipal wastes   These include biodegradable detergents and sewage – treated or untreated – as well as all the non-

The global scale of the plunder of sea life is shown graphically by this map showing the increasing area of operation of Japanese tuna-fishing fleets. If one small country has such a great national appetite, increasing conflict – of interest if not in the military sense – on the high seas is inevitable. And the oceans will be the losers.

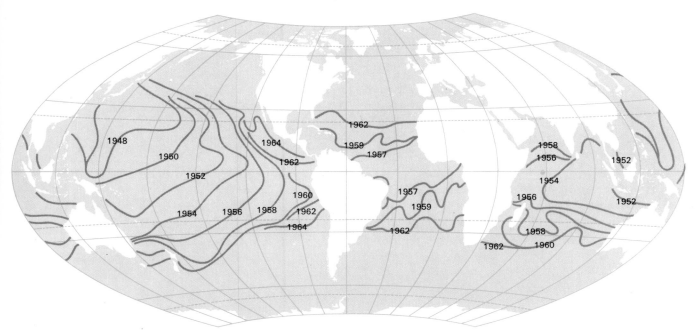

below Global shipping routes: These represent man-made oceanic circulation patterns. In contrast to fishing, merchant shipping has settled down from the age of piracy and is now subject to international rules and controls – although these are not always strict enough. These patterns are, however, subject to strong political influences; for example, war has periodically closed the Suez Canal route, shown here as open.

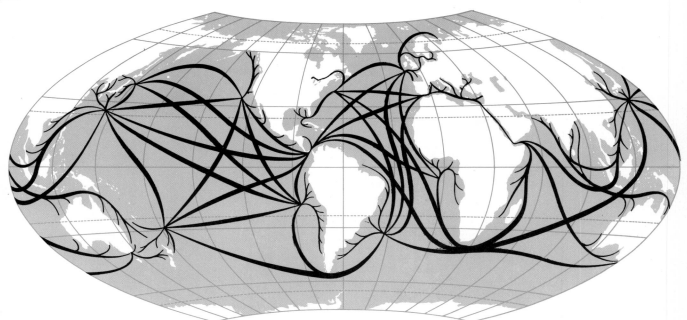

degradable garbage which is eventually strewn on our beaches. The Mediterranean countries are feeling most of all the effects of the lesser bacterial life in their sea, for modern sewage-disposal habits force the bacterial breakdown on to marine life, which is becoming increasingly unable to cope. Perhaps a return to bacterial breakdown on land is the inevitable solution.

Radioactivity This is possibly the most worrying of all the lethal by-products of human ingenuity. In the quest for military power and cheap electrical energy, mankind has created long-life 'hot' wastes. These genetic dangers accumulate along food chains and are dispersed worldwide. The canisters of radioactive wastes at the bottom of the sea are *hoped* to last the lethal lifetime of their contents. The consequences if they do not are unthinkable.

Detergents These include both household types (mostly biodegradable – that is, digestible by bacteria in sewage treatment plants) – and 'hard' industrial varieties. When used to disperse oil-slicks at sea, the effects may be more harmful than those of the oil. The nitrates and phosphates in detergents may over-stimulate the micro-life of closed waterways. Eventually the water becomes turbid, dead and rotting plant life use up all the available oxygen, and fish can no longer survive. This process is called eutrophication.

Industrial effluents Factories are often sited deliberately on an estuary because of cheap waste-disposal downstream from the populace. An enormous variety of chemical wastes may be released, very many of them non-degradable or actually poisonous. This is extremely dangerous in the life-nurturing shallows, spawning grounds of much marine life.

Carbon dioxide The effects of the enormous increase in carbon dioxide pouring into the atmosphere as a result of the burning of fossil fuels are disputed. An effect on climates is expected, but views differ. Much of the carbon dioxide is absorbed into the sea – but with what effect?

Miscellaneous wastes The oceans are treated as the planetary dustbins on an ever-increasing scale. Barges dump anything from dredged river-bottom sludge to barrels of chemical waste and cylinders of nerve gas.

Whose back-yard?

One of the major paradoxes of the oceans which keeps recurring in different guises is the fact that the oceans belong to nobody and yet to everybody. Mankind is ponderously but surely being forced into realizing that the oceanic dumping grounds are not only his neighbour's back-yard but also his own. Furthermore, the realization is dawning that far more depends on this aquatic back-yard

above Like a phoenix arising, a sunflower
flourishes in harbour debris. Perhaps in this
image of nature responding to and fighting
man's contamination of the environment
there is hope for the future.

left News agency pictures and report
show the dumping of 67 tons of nerve gas
in the Atlantic by the U.S. Army in August
1970. The dumping took place in spite of
international protests and a court action, and
probably represented the most lethal
deliberate dumping ever to have taken place
in the seas. It seemed to show a supreme
disregard for natural circulation patterns, for
the dumping site was close to the Gulf
Stream.

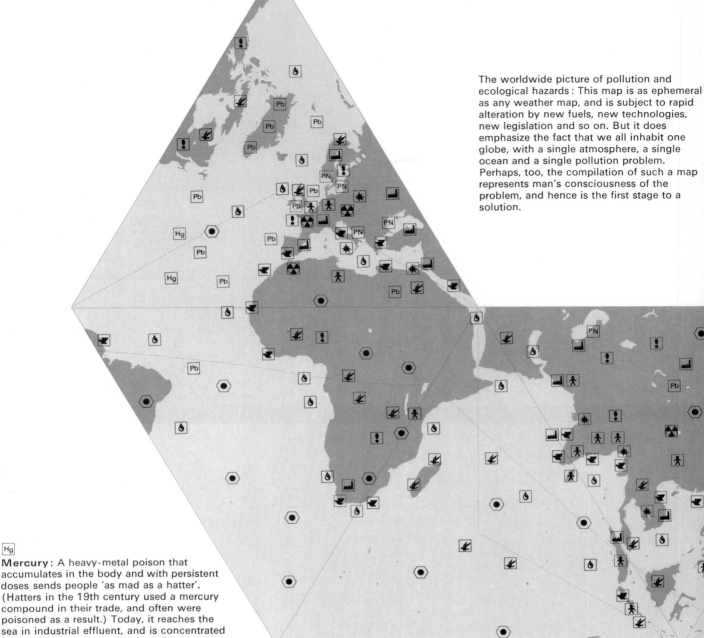

The worldwide picture of pollution and ecological hazards: This map is as ephemeral as any weather map, and is subject to rapid alteration by new fuels, new technologies, new legislation and so on. But it does emphasize the fact that we all inhabit one globe, with a single atmosphere, a single ocean and a single pollution problem. Perhaps, too, the compilation of such a map represents man's consciousness of the problem, and hence is the first stage to a solution.

Hg

Mercury: A heavy-metal poison that accumulates in the body and with persistent doses sends people 'as mad as a hatter'. (Hatters in the 19th century used a mercury compound in their trade, and often were poisoned as a result.) Today, it reaches the sea in industrial effluent, and is concentrated in the bodies of many fish, including tuna and swordfish. A number of people eating contaminated fish have died, particularly in Japan. Recently, another and closely related metal, cadmium, has been identified as a similar hazard.

Pb

Lead: Another heavy metal that collects in the body and damages the brain and nervous system. Many scientists put the major blame for lead pollution on a single source: anti-knock compounds added to car fuel to raise its octane rating. When the fuel burns, the lead passes into the atmosphere in minute particles in the exhaust fumes. The lead eventually is washed into the rivers and seas. One hope is for the increasing use of low-lead, low-octane fuels: these are necessary in cars fitted with equipment to curb other types of exhaust pollution, because lead 'poisons' the catalytic converters that break down other pollutants.

Industrial effluent: Apart from those specially mentioned, a great variety of industrial wastes are discharged from factories or dumped in the oceans. Unlike sewage,

these wastes cannot generally be broken down into harmless substances by natural process. By his thoughtless dumping, man is interfering with the whole balance of natural forces – with increasingly dire results.

Organo-chlorine compounds: Persistent man-made chemicals that include pesticides such as DDT and industrial chemicals known as PCBs. Both find their way by air and water into the seas, where they are concentrated up marine food chains. As a result, the carnivores at the head of the food chains – such as large fish and sea birds – contain in their tissues large amounts of these poisons. PCBs have been identified as the major cause of the death of some 100,000 sea birds in the Irish Sea in 1969.

PN

Phosphates and nitrates: Both ingredients of synthetic fertilizers and detergents. When washed off the land or discharged in sewage into enclosed waters, they may cause such luxuriant growth of plant life that the waters become choked with rotten, decomposing matter. The result is that the water becomes

deoxygenated, and animal life cannot survive. The Baltic Sea and the Great Lakes of North America are among the areas threatened in this way.

Oil: World reserves of petroleum are dwindling fast, but in the meantime the oceans of the world are being increasingly polluted. Prime cause is the deliberate dumping of oil residues as tankers clean their empty tanks on the high seas. More spectacular are the massive oil spills due to tanker wrecks, and oil companies increasingly exploit offshore oil fields, posing a third threat. Nevertheless, oil can be more of an aesthetic hazard than a real threat to ocean life; in fact, the detergent used to disperse oil slicks can be more harmful than the oil itself.

Disaster: Areas where man's thoughtless activities have resulted in catastrophe. The causes can be varied: poor farming methods and stripping of natural cover leading to dust bowls, massive mining operations leaving the land scarred and lifeless, chemical pollutants turning rich lakes and coastal waters into stinking aquatic deserts. Some of these disasters are irreversible with our present knowledge; others can be cured, but only with great effort and expenditure.

Danger: Potential disaster areas where the dangers are clear but avoidable. Examples include the Alaska oilfields, where uncontrolled activity could wreck a valuable natural habitat of numerous wild creatures, and many other areas where the spread of industrial civilization brings disruption alongside material benefits for men.

Population: People cause environmental catastrophes. Among the rapidly-expanding populations of developing areas there is often famine and disease, while the great urban sprawls of the giant cities create ever-growing mountains of garbage. In terms of the depletion of world resources, industrialized man poses a far greater threat than the traditional societies, for his waste of valuable materials is much greater. And so is his creation of pollutants.

Radiation: The most insidious and long-lasting of all pollutants, whether resulting from the testing of nuclear weapons or the build-up of radioactive wastes from nuclear power generation. There is no child alive on Earth today whose bones do not contain radioactive strontium-90 from the fallout following nuclear explosions. Radiation is a cause of genetic mutations: What heritage are we passing to our descendants?

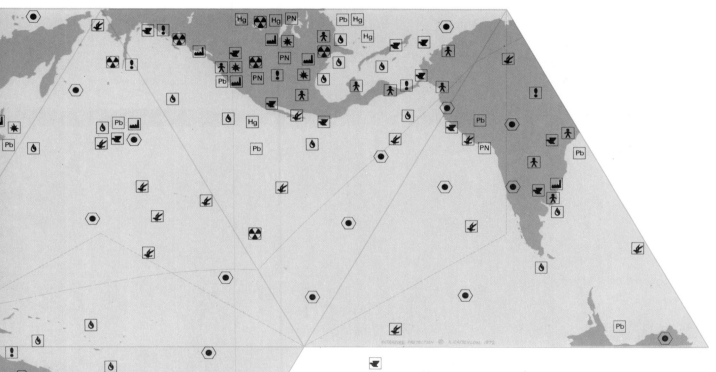

Sewage: Untreated sewage discharged into seas and rivers can pose serious health hazards and can also deoxygenate the waters, asphyxiating animal life. Sewage treatment plants can greatly reduce the problem, but they cannot cope, and even highly industrialized countries discharge large amounts of raw sewage. Sewage disposal also represents a great wastage of fresh water, a valuable natural resource that is rapidly becoming scarce. Technology exists to extract drinking water from sewage, but it is expensive. The correct residue of human sewage must find a way back to the soil whence it came to complete the cycle.

Wildlife in danger: Expansion of industrialization, removal of natural habitats, hunting and trapping, pollution, and many other factors are threatening wild creatures on every continent and ocean. The tiger and leopard are among the most spectacular threatened species, but it is also estimated that one in every ten species of plants faces extinction.

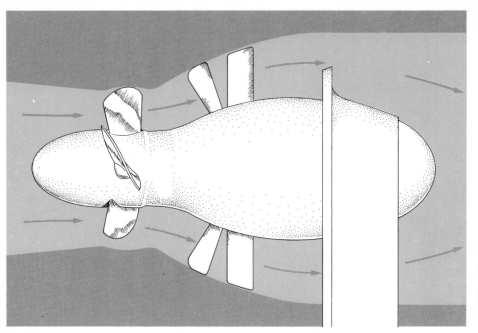

Generating power from the tides offers a clean, responsible alternative to burning up irreplaceable fossil fuels or harnessing nuclear fission, with its inevitable radioactive waste disposal problem. **left** A Russian design for a tidal turbine; **right** the Rance dam – the world's first tidal power plant – under construction near St Malo, France.

than people dreamed of – climate, food, water supply and other vital needs. Sooner or later schools will have to teach that the planet is a closed system: what goes in at one place must sooner or later reappear at another. There are so many people on Earth, and the fishing grounds are so well distributed, that what is a dumping ground is also a food-growing area. International agreement on dumping controls is not a romantic, idealistic dream but the only hope for a living oceanic future.

There is, however, an encouraging awareness among the most responsible scientists. This is shown by these words, written by marine scientists Donald Hood and Peter McRoy of the University of Alaska, in the book *Impingement of Man on the Oceans:*

There is no question but that we are at the cross-roads; we must choose our destiny. Either we plunge recklessly into a sea of known and potential hazards, with almost certain irreversible impact as the result of continued loss of air and water and ever-increasing discharge of waste loads, or we go into a carefully plotted course toward waste re-use, closed cycling, and minimal discharge to the environment except for intended non-damaging or even beneficial materials.

The latter choice is the only one tenable. The costs will be high: the advancement in standards of living (creation of wastes) must cease and the philosophy of economic gain through increased per capita consumption will have to be eliminated. Not to live compatibly with the living organism Earth is not to live at all; *the oceans are our last frontier. There is yet time to eliminate the smog, clean the rivers and estuaries, reclaim the lakes, and keep the ocean clean. But it is now only a few minutes to midnight. We on the Earth must respond now – in this decade – or we will lose our destiny.*

This is an unusually outspoken and challenging climax to the most authoritative academic volume on man's effect on the oceanic environment. There are within it, though, some very sensitive and controversial points. Hood and McRoy not only attack the very theoretical basis of current economics by dismissing per capita consumption, but also reject the advancement in standards of living, equating this with increased material possessions and energy consumption. Finally, there remains the massive difficulty of convincing the world community that those countries who have acquired these high levels of energy consumption and material standards of living can afford to call a halt. But what about those who have suffered European interference for the last 150 to 200 years, who have yet to climb over the subsistence level? It is very idealistic to expect them to agree to a halt in increased living standards. This latter question is very much at the root of the problem of agreements on fishing rights and catch limits that will not deplete world stocks of fish. The nations with greatest needs compete at a great technological disadvantage with those with the greatest wants. Much larger issues of distribution of wealth on a global scale are bound to emerge if binding, lasting and just agreement is to be established. And agreement *must* be reached, for survival is a total issue.

Chapter 5

The Last Resource?

With responsible management and proper understanding, the oceans can bring man great benefit in the future. **opposite** Fish weirs in Dahomey, west Africa, illustrate primitive but responsible harvesting of the seas. **below** A nuclear missile is launched from a submerged Polaris submarine; must this be the pattern of mankind's oceanic future? **left** Scientists monitor a spacecraft launch. Man has the technology to apply similar monitoring to the oceans; all he needs is the goodwill and faith in a group survival future.

WHO could have believed that those vast, hostile expanses of ocean water held unlimited riches? Abundant food, oil, natural gas, chemicals, minerals (including diamonds and gold), sweet water, fertilizers – all are there for man's taking. It is a rosy picture that has been painted by many – too many – authors in recent years. Where is the catch? This survey has demonstrated that there are indeed vast resources in the planetary oceans, but resources for whom? It is all too apparent that if men exploit the oceans in the way they have exploited the land since the industrial revolution, then the results will be disastrous. It is quite plain that all the major forces that go to make up our planetary home, including life, work cyclically. The oceans can be regarded as a massive source of wealth only if we recognize our debt to the system before plundering. We still have a chance to learn; in this sense, the oceans represent our last resource.

What resources?

There are several basic aspects to 'resources' – energy and materials – which have a vital bearing on man's relationship to them. The first is the distinction between 'income' and 'capital' resources. Incoming resources are primarily sunlight (much of which is impounded by the green life of the planet), cosmic rays, cosmic dust and meteorites. Capital resources are those finite elements from which the planet is built; among the special examples are the fossil fuels –

oil, natural gas and coal – which took millions of years to lay down but seem likely to be irreplaceably burnt up within the next century or so.

Secondly, it is important to check the mistake of talking too freely about the consumption and recycling of materials and energy. Both concepts are subject to the fundamental laws of matter. In the first place, the physical universe appears to us as a finite totality which cannot gain or lose energy in any form. In the second place, life gets the maximum out of its resources by a brilliant use of recycling, but this is governed by the law of entropy, and eventually the energy 'borrowed' from the system has to be returned in a form that can no longer be used. In the end it must return to space as the lowest form of energy, radiant heat. So it is incorrect to take up an entrenched position either as an incurable optimist or as a diehard pessimist when it comes to human resources. Nothing can be completely 'consumed'; it can only change its form. As far as humanity is concerned, the operative consideration is whether or not it is retrievable. Neither is anything capable of 100 per cent recycling, as some optimists have implied, although good management can achieve levels of 90 per cent in the case of some resources.

A careful consideration of both aspects is absolutely essential if men and the environment are to gain long-term value from oceanic resources. The whole concept of ecology implies the cyclic use and re-use of resources by

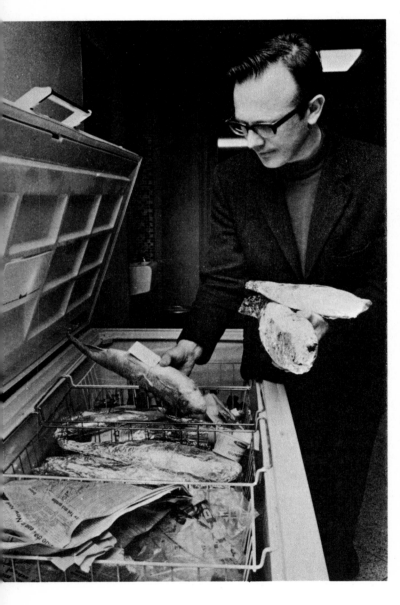

An American scientist examines fish found to contain so much mercury that they are unfit for consumption. Events such as this have brought home to some people the fact that wastes that are 'thrown away' are liable to return and plague man. But what is being done to monitor and control the discharge of all industrial effluent?

various parts of the living world. At the same time, cyclicity is run on 'consumption' or energy conversion. Life has devised a complementary situation, with the plant world 'breathing' out what animals breathe in. The vegetable world simultaneously supplies oxygen as a function of converting sunlight, as it forms food for animal life. The animals return the 'borrowed' materials to the soil as faeces and urine – or did so until we got out of step, with our post-industrial-revolution sewage ideas and intensive farming methods. In the same way, trees return leaves – which have functioned as solar energy converters during the summer months – to become soil at their own roots.

The third major factor that completes this picture of resources is human attitudes. This has become our most vital resource, for it can make or break the delicate web of life. There is no point in talking of the unlimited riches of the oceans unless there is a change in present attitudes and behaviours towards material resources. We are 'consuming' and neglecting to recycle to a suicidal degree in our present world economies. To expend the oceans' resources rather than to cultivate recycling would be not only irresponsible and uneconomic but disastrous even in the short run. As we are using up our valuable and irreplaceable fossil fuels, we must very soon turn to the 'income' energies that nurture green life and are constant and free. The other by-products of this solar bounty are the winds, rains, rivers and tidal forces – all responsible economic prospects if men's attitudes can be redirected and his ingenuity brought to bear on devising conversion mechanisms.

But what positive attitudes can be taken and what positive moves made? It seems quite without question that the best model available to mankind is the biological one, based on general individual awareness and personal responsibility. The greatest mark of biological evolution is the use of recycling and distribution. It is through recycling that life forms have changed the face of the Earth and atmosphere. It is only through sound agricultural policy that any lasting human civilization has been able to be built. And agriculture, by the very definition of the word, means a direct exchange of organic fertilizer for nurtured and harvested crops. As a dynamic equilibrium is the basis of sound ecology, it should not be startling for man to have to implement this principle as a self-conscious action.

Towards responsible marine management

If mankind is going to complete the experiment of a self-conscious creature taking guardianship over the planetary biosphere, then it is obviously essential that he goes fully

right The world fish catch has grown rapidly, keeping up with the increase in world population. In fact, the total catch trebled in the 30 years from 1938 to 1968. But such figures hide important facts. First, only half of all fish caught is consumed by man; the rest is fed to livestock, representing overall a considerable food loss. Second, the increase has taken place only at the cost of seriously depleting stocks of live fish in the oceans. **below** This map of the North Atlantic shows when various species in various waters reached the limits of productivity through over-fishing. At the dates shown, increased fishing no longer resulted in an increased catch; the fish were not breeding fast enough to replace those caught.

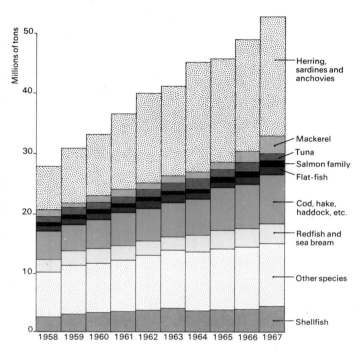

Millions of tons

Herring, sardines and anchovies

Mackerel
Tuna
Salmon family
Flat-fish

Cod, hake, haddock, etc.

Redfish and sea bream

Other species

Shellfish

1958 1959 1960 1961 1962 1963 1964 1965 1966 1967

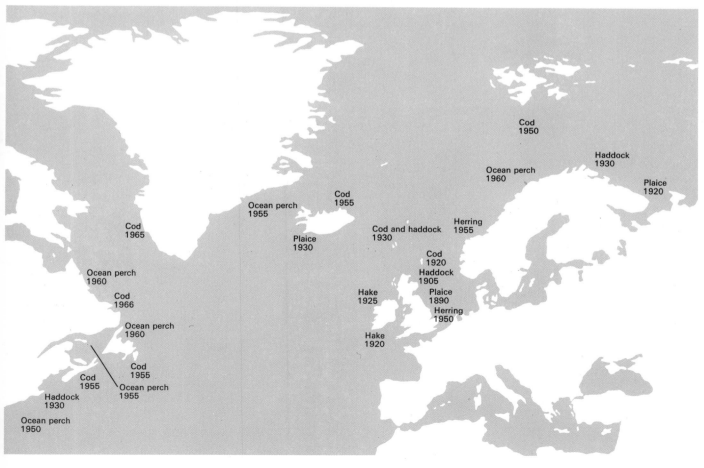

Cod
1950

Haddock
1930

Ocean perch
1960

Plaice
1920

Ocean perch
1955

Cod
1955

Cod
1965

Plaice
1930

Cod and haddock
1930

Herring
1955

Ocean perch
1960

Cod
1920

Cod
1966

Haddock
1905

Hake
1925

Plaice
1890

Ocean perch
1960

Herring
1950

Cod
1955

Hake
1920

Cod
1955

Ocean perch
1955

Haddock
1930

Ocean perch
1950

The two faces of ocean harvesting:
right The hunting and butchery of whales represents the most irresponsible type of oceanic plundering. Thanks to modern technology, the process is ruthless and efficient, but morally it belongs in a degenerate bygone age. **below** In a Japanese oyster farm, millions of oysters are reared in shallow estuarine water. Clearly, there are no problems of fencing in the 'stock', as there would be with free-swimming fish, but fish-farming must be developed in the future if the oceans are to yield the maximum food and not be denuded of life.

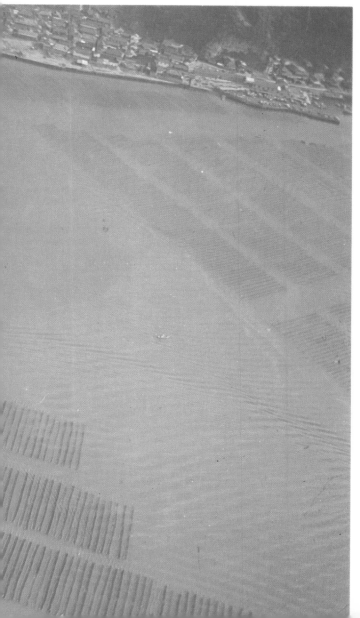

into and understands the aquatic environment. This understanding, as has been demonstrated, can no longer be on the basis of capital exploitation for the benefit of a minority of men, or even for all men to the exclusion of the rest of the biosphere. The role of the seas is paramount in a global picture, and our new attitudes will have to parallel those of the wise peasant farmer or husbandman – a sensitive modesty reinforced by the best scientific methods available.

The first principles of any husbandry must surely be to know one's 'soil', one's primary vegetation, one's climatic behaviour and the various kinds of grazers or feeders. According to which stage one is concerned with in the cultivation of living systems, each 'consumer' can be regarded as hostile or beneficial. The use of pesticides in the growing of grain crops is based on the production of grain for human consumption. If the aim were to cultivate wild birds who prey on the so-called pests, then those very pests would be encouraged and not exterminated. It seems obvious that mariculture should not begin with a one-crop system, as this has blighted too many 'scientifically'-orientated terrestrial patterns. Harvesting will have to be on a 'filter' basis – that is, the aim will have to be aquatic environments accommodating many species, and man will have to carefully determine how much he can safely filter out of the culture, based on a total monitoring of what he puts in. It seems possible and even desirable that aquaculture could have as near to a 'wild park' atmosphere as man could devise. This would be quite contrary to the principle of present-day intensive egg and meat farming.

If ocean farming is proposed, on what basis can one expect agreement between nations that have so long warred over such things as fishing rights? One only has to check the records of fisheries and whaling catches to see the overkill effects of unlimited competition. The blue whale catch in 1930–31, for example, was 80,000; in 1963–64 only 372. The Pacific sardine catch improved from 25,000 tons in 1916 up to 790,000 tons in 1936, only to rocket down to 40,000 tons in the 1960s – a drop back to one eighteenth the pre-war overkill! The answer is shown in the statistics of the Pacific halibut catch. Between 1915 and 1930 it dropped by two-thirds. Foresight impelled a treaty enforced by 1932. The 1955 halibut catch in this region was the highest recorded, at four times the 1930 total. A free-for-all has consistently ended in a none-for-all. The sooner fishing nations take even the elementary step of treaty enforcement the sooner will we have fishing stock to even speak about. Enforcement will always be a thorny

Sketches by the author establish a rational approach to the problem of designing a fish-farming complex that is integrated with its environment. It must be emphasized that these are diagrammatic explorations of a proposed design approach to aquaculture; the text discusses in more detail the overall ethos of ocean farming.

1 The two fundamental behaviour patterns of vortices in water: In the first instance, a stream of water enters still water through a narrow opening, and the vortices spin outwards from the flow. In the second, the water is moving and the vortex train is caused by an obstruction, which diverts the spin inwards.

2 The integration of fish propulsion and natural vortical behaviour of the disturbed water: The arrows indicate the directions of water particles. The centres of the eddies act as 'support posts' for the sides of the fish to push against.

3 An analysis of the vortex trains according to the proportion of height to length of pitch: Three types are shown, with the smaller representing a faster movement and the larger a slower movement. The proportions are specifically chosen as a basis for the design of volumes and water-'field' boundaries. By following the natural 'fields'

of water flow (which also operate on a scale of miles), the controlled fields for filtering inhabitants would give maximum natural distribution and cause minimal interference with the environment.

4 A comparison of two types of eddy arrangements: The drawings show three factors: firstly, the path of a body moving through water; secondly, the direction of the water each side of this body moving back into the space it has vacated (larger arrows); thirdly, the direction of the water particles (smaller arrows). The upper diagram shows so-called shifted eddies, where the centres of rotation of the particles of water are not in parallel. The lower diagram shows a straight line of eddies, where the water particles develop symmetrical and balanced vortices.

5 The development (from left to right) of harnessing a vortex pattern and rationalizing it into a geometrically ordered pattern. It can be seen that this particular natural pattern lends itself to analysis into an overlapping hexagonal module where the edges of the hexagons rest on each other's internal triangle.

6 Detailed rationalization of the hexagonal vortex 'fields'. The whole system can be reduced to a common triangular component. Each hexagon is composed of 96 such triangles, with a four-triangle edge module and a six-triangle edge length to the internal triangle. Where they overlap, each pair of hexagons shares 16 triangles. From this triangular grid, it is easy to calculate the exact surface area of any part. Also, the triangular net would make an ideal form-controlling device for the boundaries of the fish-farm 'fields'. The sets of five shaded triangles (subdivided into smaller triangles on the left) represent the best monitoring areas for both 'interior' waters and through-flow channels.

7 A diagrammatic layout attempting to gain maximum advantage from the vortical-flow principle while integrating it with area-controlled close-packing, so that no space is wasted. Here 'fields' can follow a 'linear' or 'surface-area' pattern, according to need and management. There are regularly-spaced centres that can control or monitor four adjacent oceanic 'fields'. Lines indicate main communication routes, while the shape on the right has been divided to demonstrate the squares that control its surface area.

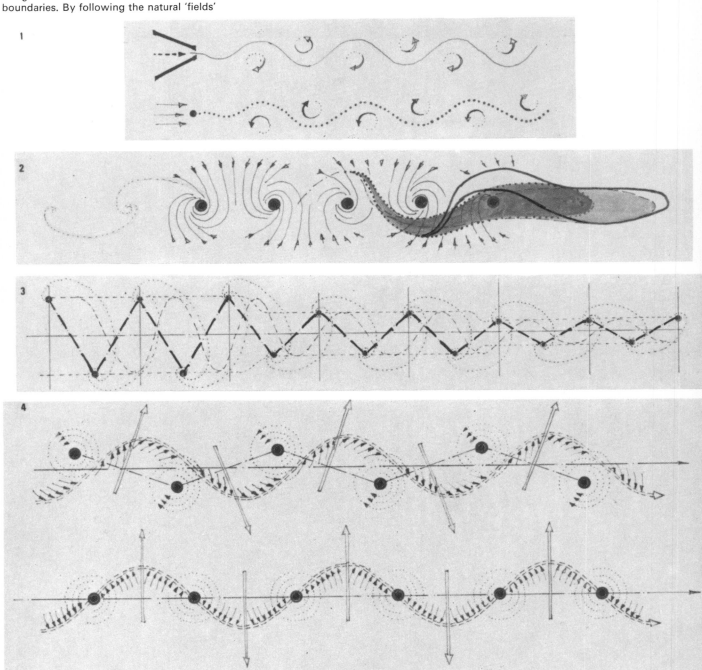

above In a highly mechanized plant — the equivalent in some ways of a factory-farm — Pacific prawns are reared and grown in Japan. Their conditions are scientifically monitored and carefully controlled, and millions of the prawns are exported.

right Floating aquaculture is visualized in the Gulf of Timor or similar equatorial waters, where existing islands can be utilized to relate to the various fish-farming communities and their floating fields. These speculative proposals are based on the principle of polyculture, reproducing a balance of species as near to the wild as possible and filtering off the excess produce with minimal interference. In the right foreground is a solar distillation plant for obtaining fresh water that is visualized as being capable of growing grain and land plants by hydroponic methods. The field system is based on a curvilinear symmetry (see plan inset *right*) which is a balance between the vortical flow of the liquid environment and an area division that can be precisely calculated. Many different field systems could be used to give variety to the marine environment. Many kinds of floating structure are envisaged, their design basis being a cross between a buoy and a boat.

IN CHOOSING POTENTIAL SITES FOR SWEET AND SALT WATER AQUACULTURE ON A GLOBAL SCALE CERTAIN CONDITIONS RECOMMEND THEMSELVES. FIRSTLY THE VAST QUANTITIES OF RAIN-WATER FALLING INTO THE EQUATORIAL SEAS. SECONDLY THE STRONGEST SUNSHINE AREA WILL ALSO BE EQUATORIAL, ALTHOUGH NOT NECESSARILY THE LONGEST SUNSHINE HOURS. THIRDLY THE SAME EQUATORIAL BAND HAS GENERALLY THE LEAST WIND. THIS IS BECAUSE OF THE NATURE OF THE WINDS GENERATED IN BOTH NORTH AND SOUTH HEMISPHERES.

WHEN THEY MEET IN THE CENTRAL AREA OF THE GLOBE THEY TEND TO CANCEL EACH OTHER OUT. THESE AREAS OF CALM ARE TRADITIONALLY KNOWN AS THE DOLDRUMS OR CALMS. THEY WERE DISCOVERED BY THE GLOBE TROTTING SAILING SHIPS AND WERE THE CAUSE OF MANY AN UPSET SCHEDULE. THIS PAGE CHOOSES A HYPOTHETICAL SITE AND EXPLORES CERTAIN OF THE DESIGN CRITERIA.

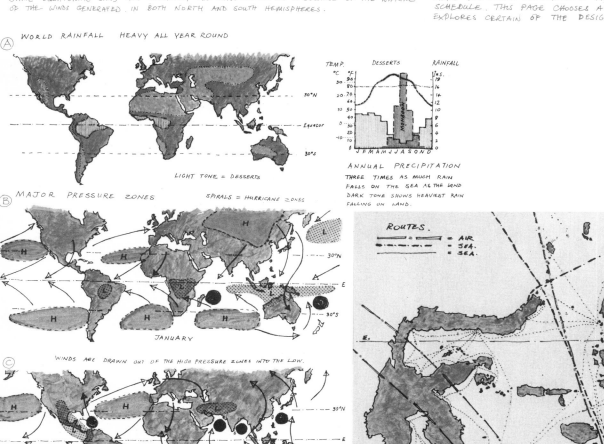

(A) WORLD RAINFALL   HEAVY ALL YEAR ROUND

LIGHT TONE = DESSERTS

TEMP.   DESSERTS   RAINFALL
°C °F   INS.

ANNUAL PRECIPITATION
THREE TIMES AS MUCH RAIN FALLS ON THE SEA AS THE LAND. DARK TONE SHOWS HEAVIEST RAIN FALLING ON LAND.

(B) MAJOR PRESSURE ZONES   SPIRALS = HURRICANE ZONES

30°N
E   E
30°S
JANUARY

WINDS ARE DRAWN OUT OF THE HIGH PRESSURE ZONES INTO THE LOW.

(C)
30°N
E
30°S
JULY

CIRCULAR SPIRALS = HURRICANE ZONES.   HIGH PRESSURE ZONES DONATE WINDS.

30°N   30°S

(D)

DARK BAND REPRESENTS THE 300 MILE BROAD SUNLIT CALM WATERS WHICH HOLD POTENTIAL FOR AQUACULTURE AND POWER CONVERSION.

(A)(B)(C)(D) SHOW FOUR PROJECTIONS OF THE WORLD DEMONSTRATING FACTORS WITH DIRECT BEARING ON AQUACULTURE. TOP (A) ANNUAL RAINFALL (B) JANUARY HIGH PRESSURE ZONES WHICH DONATE WINDS. THE CIRCULAR SPIRALS ARE ON THE HURRICANE SITES WITH THE ARROWS INDICATING WIND TRENDS. (C) THE JULY PATTERN (D) DEMONSTRATES THE EQUATORIAL BELT WITH ITS GREAT PROPORTION OF THE SEA SURFACE.

ROUTES.
= AIR
= SEA
= SEA

130°
130°

60°
120°   130°   140°
0°
50°
120°
SCALE   MILES

THE BRITISH ISLES AND CELEBES (SULAWESI) COMPARED IN SIZE BUT NOT IN THE SAME LONGITUDE.

THE DRAWING ON THE RIGHT IS A DIAGRAMMATIC VERSION OF THE PRINCIPLES OF INTERACTION BETWEEN AIR MOVEMENT AND WATER. THE ACTION CREATES WAVE MOTION ON THE ONE HAND THROUGH THE ROTATION OF THE WATER PARTICLES, AND ON THE OTHER HAND A DEEP SPIRAL IN THE WATER KNOWN AS THE EKMAN EFFECT. THIS IS SHOWN BY THE SMALLER ARROWS IN THE DIAGRAM. THE ACTUAL EFFECTS ARE MULTIFARIOUS COMPLEX AND BEAUTIFULLY RELATED. THESE COMBINED INTER-ACTIONS ARE VITAL CONTRIBUTORS TO THE DIFFUSION AND DISTRIBUTION OF THE ELEMENTS OF THE AQUATIC ENVIRONMENT.

THE LOWER HALF OF THE DIAGRAM DEMONSTRATES THE NATURE OF THE 'ROLLING' PARTICLES OF WATER AS THE WAVE PASSES THROUGH THEM. THE ROTATIONS DIMINISH WITH DEPTH AND HAS A CHARACTERISTIC EIGHT FOLD NATURE TO ITS OSCILLATION. THE EKMAN SPIRAL ALSO DIMINISHES WITH DEPTH AND THE 'TURN' ITSELF IS AN EXPRESSION OF PRECESSION.

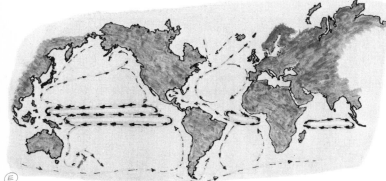

(E) ANOTHER GENERALIZED WORLD MAP SHOWING THE MAJOR WORLD SURFACE OCEAN CURRENTS. ALL HAVE CIRCULATIONS INDIAN, ATLANTIC AND PACIFIC THE LATTER HAVING A DOUBLE LOOP. THESE CURRENTS SUGGEST NOMADIC AQUA-CULTURE. FLOATING FARMS COULD TRAVEL AROUND THESE ROUTES AT THE TWO OR THREE YEAR PACE OF THE CIRCULATIONS THEMSELVES.

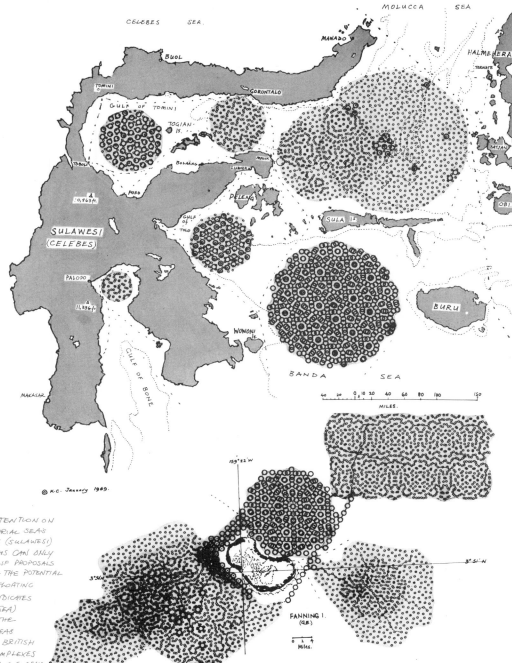

MOLUCCA SEA

CELEBES SEA.

MANADO

HALMEHERA

TERNATE

BUOL

GORONTALO

TOMINI

GULF OF TOMINI

TOGIAN IS.

BOLAÄNG

LUWUM

MALIK

BATJAN

TOBOLI

POSO

PELENG

OBI.

SULA IS.

SULAWESI (CELEBES)

△ 10,963 ft.

GULF OF TOLO

PALOPO

△ 11,336 ft.

WOWONI IS.

BURU

MAKASAR

GULF OF BONE

BANDA SEA

40 20 0 10 20 40 60 80 100   150
MILES.

© K.C. January 1969.

159°22' W

3°51' N

3°51' S

FANNING I. (G.B.)

0 2 4
Miles.

THE AREA CHOSEN TO FOCUS ATTENTION ON
IS THAT PORTION OF THE EQUATORIAL SEAS
THAT STRETCH BETWEEN CELEBES (SULAWESI)
AND THE MOLUCCAS. ALTHOUGH THIS CAN ONLY
BE CONSIDERED AN OUTLINE SET OF PROPOSALS
SOME IDEA CAN BE INDICATED OF THE POTENTIAL
SCALE OF HARVESTABLE AREAS AND FLOATING
'FARMS'. THE DRAWING ABOVE LEFT INDICATES
PRESENT TRANSPORT ROUTES. (AIR and SEA)
THE LOWER DRAWING (LEFT) INDICATES THE
SIZE AND SCALE THE PROPOSED AREAS
WOULD BE IN COMPARISON TO THE BRITISH
ISLES JUST ONE OF THE LARGER COMPLEXES
COULD YIELD A GREATER HARVESTABLE SURFACE
AREA THAN ALL THE LAND PRESENTLY
FARMED IN THE UNITED KINGDOM.
MANY KINDS OF CULTURE ARE ENVISIONED
GROWING OUT OF EXPERIENCE AND
EXPERIMENTATION.

THIS LARGER CLOSE-UP' OF THE AQUACULTURE FARMS (ABOVE) TAKES TWO TYPES
OF EQUATORIAL SITUATION: THE FIRST IN RELATION TO GULFS AND ISLAND GROUPS
IN THE MOLUCCAS AND BANDA SEAS THE SECOND LOWER COMPLEX IS ATTACHED
TO A CORAL ISLAND (FANNING) AND FREELY FLOATING AROUND IT.
IN THE LARGER COMPLEXES ABOVE IN THE GULFS OF TOMINI AND BONE AND THE
SEAS OF BANDA AND MOLUCCA, DIFFERENT KINDS OF 'PATTERN' ARE PROPOSED WHICH
ARE BASICALLY ANCHORED BUT WITH THE FUNDAMENTAL ETHOS OF LEAST INTERFERENCE
WITH THE NATURAL ORDER, HARVEST BY FILTRATION AND MAXIMUM VARIETY OF HUMAN
SETTLEMENT TYPES. EXPERIMENTATION WITH HYDROPONIC CREEPER AND SHRUBS
WOULD BE ENCOURAGED TOGETHER WITH POSSIBLE HYDROPONIC RICE PADDIES. THE
STRUCTURES ARE CONCEIVED AS 'GROWING INTO' THE PROPOSED AREAS AS FARMS, FARMERS
METHODS AND ECONOMIC SYSTEMS DEVELOP. IT WOULD NOT BE CONCEIVED IN ANY
WAY AS A 'MASTER' PLAN BUT AN ORGANICALLY RESPONSIVE PROGRAMME
WHICH WOULD BE SENSITIVE TO HUMAN AND ECOLOGICAL NEEDS AS THEY DEVELOPED
AND WOULD BE CAPABLE OF CONTRACTION IF ENVIRONMENTAL RESPONSES WERE
NEGATIVE.
NOT ONLY COULD SUCH AREAS PRODUCE STAPLE FOODS FOR THE GREATER POPULATION
PRESSURE AREAS BUT MIGHT EVEN PROVIDE REFUGE FOR THE MANY TRADITIONAL
SOCIETIES AT PRESENT THREATENED WITH EXTINCTION AS WELL AS AREAS THAT
COULD PROVIDE THE MEANS FOR NEW SOCIAL PATTERNS AND COMMUNITIES.

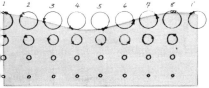

THREE AND TWO DIMENSIONAL DIAGRAMS OF WATER BEHAVIOUR
IN RESPONSE TO ATMOSPHERIC STIMULI.

problem, based ultimately on enlightened self-interest, but some sort of collective policing will have to emerge as surely as the high sea pirates were eventually defeated. Mariculture of the future will plan to avoid ignorant over-exploitation based on ill-informed estimates of the oceanic environment's carrying capacity.

The shape of marine farming

Aquaculture or mariculture is no modern dream. Laws regulating fisheries were in operation in both Sumeria in 2000 B.C. and in China 2000 years ago. It has been a fact for centuries among the Pacific islands, and has been developed to perfection in many aspects over a similar

time-span by the Japanese. The question today is how to expand these arts, maintaining a scale that will encourage 'free' local participation, as well as developing larger-scale internationally-monitored farms. What are the technical difficulties? And what do the experts say? In taking advice from experts, we have to steer a careful course between academic enthusiasm and economic reality. The key to predictions and projections is to make sure one's time-scales are in reasonable order, and in assessing economics it is imperative to take into account the costs of *not* taking timely initiative.

Timothy Joyner of the U.S. National Marine Fisheries Service will be first spokesman. Land farming, he argues, not only selects and nurtures organisms beyond their natural survival but also releases men to produce the flowering of cultures. The limits of man's sea harvesting are much harder to assess. What can be observed in history is a pattern of extended fishing powers becoming one of the bases of man's political power and commercial empire-building – and eventually the abuse and overkill of the resource on which the power was built. Paradoxically, industrialization – including the building of docking facilities and the production of industrial wastes – soon disrupts the very spawning grounds that the fleet is designed to plunder. The nineteenth century brought simultaneous awareness of overkill abilities and conservation necessities based on data won by the same ingenuity. The recurrent problem arises – how to occupy and use a range without destroying its ability to sustain its occupants.

The three judicial zones of the hydrosphere, Joyner goes on to argue, have always been the internal waters – rivers and lakes – of a nation, its territorial seas, and the high seas that come under no nation's sovereignty. Two new zones emerge from increased pressure: the contiguous zones on

above A design for a floating wind-powered generating station is based on a land design 140 feet high. It has hollow propeller blades. If the technology were developed fully, a chain of such generators on land and sea could supply much power for mankind. **right** The areas of the globe with the greatest wind power potential are shaded. **opposite** A scheme by the author combines power generation by nuclear fusion, the distillation of fresh water from the sea and the extraction of minerals from sea water. Only speculation to stimulate research at present, such a centre could supply many needs of nearby floating communities.

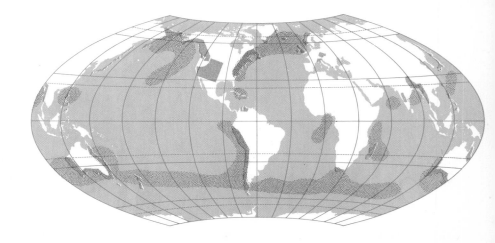

Communities do exist who spend virtually their whole lives on the water, and we can learn much from them of the problems facing the inhabitants of floating villages in the future. Motives for living on the water vary widely. **above** For the Orang Laut of the Malay Archipelago, the sea is their traditional home. **above right** For the inhabitants of the sampans and junks of Hong Kong, the reasons are largely economic and social. They generally cannot afford to live on land, or cannot find homes in the overpopulated colony. **right** The crew of an aircraft carrier are brought together to form a sophisticated community for military reasons reflecting today's power politics.

the edges of national boundaries and the continental shelves. In these zones it is reasonable to expect vigorous competition for resources between such interests as aquaculture, commercial fisheries, recreation, waste disposal, shipping, petroleum production, marine mining and electric power generation. With this realism in mind, Joyner suggests that aquaculture can be greatly expanded. Mussel culture, he points out, is vastly under-utilized, as is underwater aquaculture – sea-bottom ranching and submerged off-bottom suspended systems of shellfish cultivation – making possible, in Joyner's own words, 'truly prodigious aquacultural production'. The maintenance of 'wild' ecologies close by cultured ones, plus the imperative of avoiding single-species cultures, round off his recommendations.

The next three experts to give their findings and recommendations will be O. A. Roels, R. D. Gerard and A. W. H. Be of the Lamont-Doherty Geological Observatory, New York. They have a scheme under way on the north shore of St. Croix Island in the Caribbean, which will demonstrate the mechanical feasibility of combining aquaculture with deep-sea nutrient upwelling (by artificial pumping). This scheme could further be linked with the preparation of fresh water, using the cold deep water to condense atmospheric moisture. This cooling facility could extend to air-conditioning, conventional desalination techniques, ice-making, cooling of nuclear reactors, etc. They plan also to examine the feasibility of using the temperature differences of the deep and surface waters for generating electric power, calculating that the electricity produced could run the pump bringing up the deep-water nutrients and could also conceivably be linked with deep-sea mining operations. This joint proximity would encourage responsible cooperative functioning. Their calculations are impressive; their results will be of interest to many national and international bodies. Admiral Paulo de Castro Moreira da Silva of the Brazilian Marine Fisheries Institute has demonstrated his own ingenious methods of coordinating an upwelling (pumping) process with the production of ice (a valuable commodity in Brazil), the establishment of fertile, controlled bays for aquaculture, and the production of fresh water and salts by evaporating sea water. Again his economics are precise and persuasive.

Our next spokesman, S. J. Holt, stated his opinion based on world trends in a recent *Scientific American* article. He sees the oceans as a food source for man and his domestic animals, and as a supply of materials and medicines. But science, industry and government must work together, with

Schemes for ocean communities have great variety, according to size, materials, purpose and area of operation. **right** A British design for a structure called Sea City — partly floating and partly built on piles — has been worked out in great detail. Being designed for the North Sea, it is inward-looking and protected from wind and weather. **below** In contrast, an exploratory design sketch by the author shows a floating structure based on much lighter-weight structures. It is conceived for tropical waters and parts can separate and agglomerate according to need.

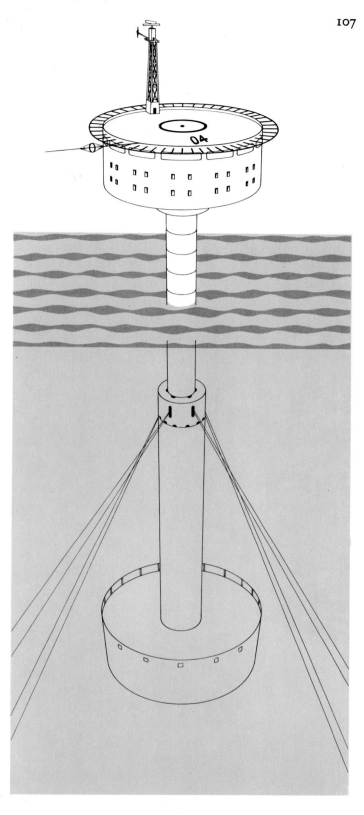

a joint ethic, without delay. Of the present sea harvest of 55 million tons, 75 per cent is taken by 14 countries. From 1850 to 1950, the world fish catch doubled and has increased even faster since. It is the only food source that has kept ahead of world human population growth. Fisheries still supply only one-tenth of man's animal protein, however, and the Indian Ocean is as yet relatively unexploited. But within ten years, Holt believes, no fishing stock will be underfished and most will be well overfished. Like the other experts, he sees the only solution in types of mariculture. Holt also believes that there are immense possibilities in mangrove swamp culture, and in breeding homing species like the salmon. The control of predators and competitors will also increase yield. He finishes with the prediction that by the end of the century the sea will be less a wilderness and more cultivated, and that it is the duty of us and our children to make sure it does not become a contaminated wilderness or battlefield between nations.

The aqua-farmers

What lessons do we have to guide us in understanding the social implications of mariculture on a large scale? Are there people who live on the floating surface of a sea environment that would give an indication that people could happily live in such a way for prolonged periods? Two such groups come to mind, apart from the traditional lake-dwelling peoples and those whose houses are built on stilts over water. These are the inhabitants of the boats in Hong Kong harbour and the people known as the Malaysian water gypsies. These latter live most of their days out to sea, just within sight of land. They live off a predominantly fish diet and come into harbour when storms threaten, on an instinctive early-warning system. The other types of people that are afloat for so much of their lives are the men who manage the world's shipping – the sailors, mechanics, technicians, and so on.

An aspect of developing mariculture that has received less attention is the possible decentralization of some of the world population by encouraging settlements in certain tropical-water development areas. Roels, Gerard and Be have pointed out that there are at least 43 extensive oceanic sites, all averaging areas as large as Britain, which have deep water close to the shore and where humid winds prevail. Here, mariculture and fresh-water production could be linked with deep ocean mining, power production, etc, and give work and homes to communities of people.

What is certain is that the development of extensive maricultural activities needs to take place as an organically

A design for a moored floating platform is rather like a giant fishing float. It is buoyant but is stabilized by underwater ballast. Such a man-made island would be stable even in heavy seas, and could be used as a mid-ocean monitoring centre. The top deck has a helicopter landing platform, and the structure could also give access to the underwater through the central tube.

based system, and should not reproduce the human error of overplanning and the blue-printing of a social system. It would ideally be a minimal structure to protect the maximum freedoms. Much the most desirable development would be the encouragement of groups of varied sizes working under a voluntary economic system which was internationally guided only by the basic ecological principles of non-overexploitation. What is most likely, however, is that the political ideology of a maricultural community will be dictated by the world area or nation that supplies the original capital. One can only hope that by this time all systems can be absorbed in mutual coexistence, except for those that importune their neighbours or transgress the biological law: What is taken out must be equally returned, in an acceptable way to the environment.

In setting up communities for aquaculture, it is essential to avoid alienating certain people to life in a barren and monotonous environment. Timothy Joyner discusses this point very sensitively. He observes that overemphasis on monoculture (single-species cultivation), resulting in loss of diversity, leads to inefficient use of the available area, to ecological imbalance that can make the crop susceptible to sudden destruction, and to a disruption of nutrient cycling and thus 'exhaustion' of the waters. All these effects can be seen in intensive land farming, and Joyner argues that they can be overcome in mariculture by simultaneous cultivation of compatible, non-competing species in each plot or enclosure, so that each species takes advantage of living conditions neglected by the others.

As an example, a single enclosure could include edible bottom-living crustaceans feeding on bottom detritus, fish and filter-feeding bivalve molluscs feeding on organisms of various sizes in the water above the bottom, and abalone or limpets feeding on the algae on the enclosure's walls. There could be great advantage in introducing land- or marsh-based vegetation, possibly grown hydroponically with water desalinated by evaporation. These plants could be a source of food as well as enhancing the visual environment of the aquafarmers with familiar land vegetation. Creepers growing on netting could not only create shade and shelter but even supply perfumed flowers to further enhance the quality of the environment.

The structures that would best serve men in an aquatic situation would also have to obey the rules of diversity. On the smaller scale, transport vehicles would be encouraged to use clean wind power. On the medium scale, structures could be designed to be able to float apart and together as the needs arose. And on the larger scale, they

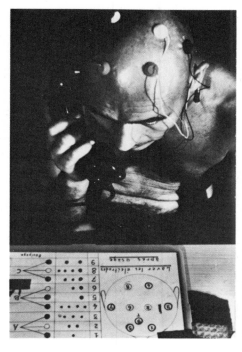

could possibly grow into a bigger variety of floating platforms with multiple deck arrangements to suit the particular needs and social structure.

These should be determined from within the floating community. Man seems to have flourished best when left alone within a simple set of vital rules and encouraged to freely explore variety of expression within these. The sophisticated factors in the aquacultural centres would be the most modern monitoring devices and oceanographic and biological laboratory centres. But in other respects they offer a unique opportunity for a move away from increasingly complex bureaucracy, back towards the simpler sociological structure of a self-governing village community.

Adapting to the underwater environment

Man has a deep-rooted biological need to explore the unknown, and in order to carry out his exploration successfully it is necessary for him to understand his own limitations, physical and mental, to adapt himself to the constraints of the new environment, and to be prepared for the unforeseen. Before men are trained to live and work underwater, psychologists must try to understand the reasons the would-be oceanauts have for volunteering. From this, it can be decided whether or not the volunteer will be useful on a particular experiment. Obviously, no matter how good a swimmer a person is, if he is overbearing and uncooperative he would be useless as a member of a diving team whose lives will depend on sympathy and understanding of each other's work and difficulties.

Assuming the psychological fitness of a potential oceanaut, he must then undergo a very thorough medical examination. Any problems with his lungs or heart must be carefully evaluated, together with the state of his ears, nose and teeth. Bad fillings can cause terrible pain when divers return to the surface. The diver's ears are always exposed to the surrounding water pressure, otherwise the change in pressure would burst the eardrums. Divers hold their nose and blow to 'clear' their ears and equalize pressure in the same way and for the same reason as people do at take-off in aircraft.

Oceanauts, fortunately, do not have to be supermen physically or mentally, but naturally it helps if they can do things on their own particularly well. The best divers are not Olympic gold-medallists or Nobel prizewinners, but calm-minded people who do not easily break into panic when things go wrong – as they are apt to do when diving. Girls are by no means excluded from experimental diving, and have proved to be every bit as good as the all-male

above Tests conducted by Jacques Cousteau during the *Conshelf* experiments included the monitoring of brain activity and other bodily responses to the stress and pressure of underwater living. **below** This model of an underwater habitat by Dräger represents the sophisticated second generation of undersea laboratories.

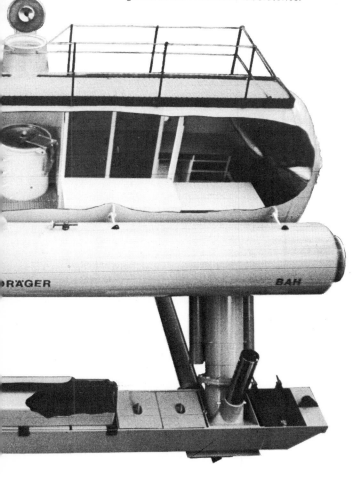

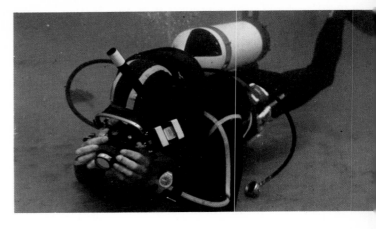

Oil	■	Gas	■
Sulphur	□	Heavy minerals	▲
Coal	▲	Tin	△
Iron	△	Diamonds	◆
Magnesium	●	Fresh water	●
Other minerals	○	Salt	○

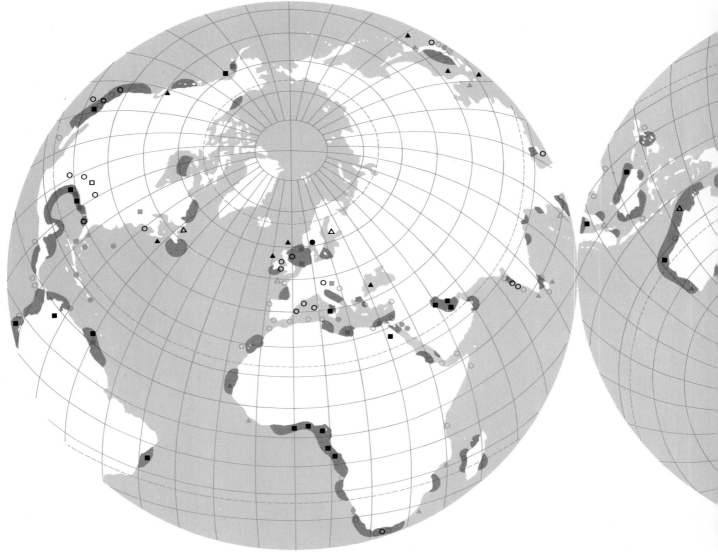

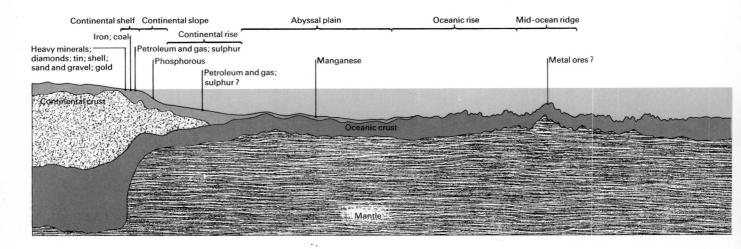

Continental shelf   Continental slope   Abyssal plain   Oceanic rise   Mid-ocean ridge

Iron; coal   Continental rise

Heavy minerals;
diamonds; tin; shell;   Petroleum and gas; sulphur
sand and gravel; gold   Phosphorous   Manganese   Metal ores ?

Petroleum and gas;
sulphur ?

Continental crust

Oceanic crust

Mantle

left In shallow continental shelf waters, geologist-divers can survey the sea bed directly, looking for signs of valuable mineral deposits. In deeper waters, surveying must be by remote control, through drilling, seismic studies and other techniques.

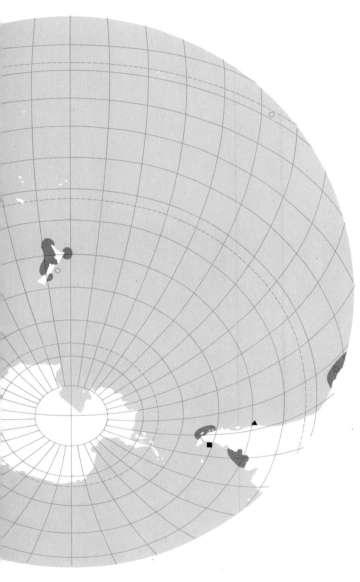

above Where man is already exploiting mineral resources in and around the oceans: Having sucked dry many land-based sources of minerals, man is expanding into the sea. But this is only putting off the day when consumption patterns will have to change. The ultimate solution to the resources problem is a change in attitude to resources. We must change from the consumption of resources to their use and re-use by recycling. **left** Increasing knowledge of the structure and composition of the Earth's crust under the oceans enables geologists to predict new sources of minerals.

teams. In September 1970 an all-girl team of divers spent 3 weeks at a depth of 60 feet in the Caribbean, living and working from a habitat called *Tektite II*.

Because of the enormous dangers, very deep training 'dives' are carried out in simulators on land before going to the open ocean. The simulators are a series of interconnecting pressure chambers which are divided into living and working compartments. One of the chambers is filled to a reasonable depth with water, so that every aspect of the hazards an oceanaut will face in the open sea can be simulated. Divers are locked into the system of chambers, and the internal pressure is increased to correspond with the appropriate depth of the dive. Every pressure increase of one atmosphere corresponds to an increase in depth of about 30 feet. Obviously an oceanaut must be gradually exposed to such a dramatic change.

The increased pressure greatly affects breathing, and special gas mixtures must be used for deep dives. Under increased pressure, gas is compressed into a smaller volume, so a diver breathes much more gas than he would at the surface. But an even more complicated problem is that too much or too little oxygen under pressure is fatal, and that nitrogen becomes narcotic at a depth of about 150 feet. Therefore, since air contains about 21 per cent oxygen and 78 per cent nitrogen by volume, compressed air can be used only for shallow dives. For deeper work, an artificial atmosphere is needed, in which a gas of a lighter molecular weight is substituted for the nitrogen. Several lives were lost in the search for a suitable inert gas. Argon, neon and others were tried. Helium and hydrogen proved the most successful, but nobody likes to use a hydrogen/oxygen mixture under pressure as it also makes a very good bomb! Today, a helium/oxygen mixture is used for deep diving.

Having been gradually compressed to several times normal atmospheric pressure in a deep dive, oceanauts must be similarly decompressed, otherwise all the gases dissolved in the blood will bubble out at once and cause decompression sickness – the 'bends'. What happens is easily demonstrated by shaking a bottle of fizzy drink before opening it; most of the drink is lost as foam. Exactly the same thing happens to the blood of a diver with decompression sickness; it can kill or permanently paralyse. The only cure is to quickly recompress and begin again.

The pull of oceanic resources

Modern man's appetite for certain physical 'resources' has become one of his greatest incentives for going into the sea. But as industrialized man benefits from his ingenuity by increases in life expectancy and leisure time, he also discovers that he is polluting the very source of his wealth. He uses as much as 25,000 gallons of water to produce one ton of steel, and twice this amount to make one ton of paper. As the demand for commodities increases, so does the demand for water. In 1900, the United States used 40,000 million gallons of water per day; by the year 2000 the demand is expected to reach nearly 1,000,000 million gallons daily. But, on the other hand, compare this with the falling cost of converting salt water into fresh: From the mid-1950s to the mid-1960s, there was a drop from five dollars per 1,000 gallons to one dollar; by the early 1970s the cost had fallen to one-third of a dollar.

As a by-product of desalination, large amounts of dissolved minerals could become available. More than 50 identifiable chemical elements occur in sea water. It has been calculated that in one cubic mile there are, among others, almost 90 million tons of chlorine, 50 million tons of sodium, $6\frac{1}{2}$ million tons of magnesium, over $1\frac{3}{4}$ million tons each of calcium and potassium, 300,000 tons of bromine, and smaller amounts of dozens of other metals, including a few pounds of gold. Since there are some 317 million cubic miles of sea water in the oceans, it is easy to see the double incentive for distilling fresh water and extracting minerals. It should also be mentioned that much oceanography in recent years has been carried out in search of minerals and fossil fuels on and under the sea bed. As men have exhausted irreplaceable supplies on land, they have had to go into the continental shelves to keep up with the appetites of unrestrained industrial societies. More than a dozen mining industries are now operating within the ocean fringes, mostly within major fishing grounds.

The oceans could also play an extremely important role in the production of power. As the exhaustion of fossil chemical fuels accelerates, nuclear power has been introduced to meet our huge demands. So far, this has meant nuclear fission, and however man tries to reduce the radioactive wastes, they continue to accumulate and embarrass him, threatening to poison the ocean environment. The great hope of modern physics is the harnessing of nuclear fusion – the 'clean' power-source of the Sun. By its very nature, fusion avoids the radioactive pollution of fission, and when it becomes a reality the sea's waters will become the planet's major fuel resource. It contains in its water molecules more than enough hydrogen to supply ample energy for the remaining life of the Solar System.

But if man fails in his bid for fusion power, quite revolutionary efforts will have to be directed into harnessing

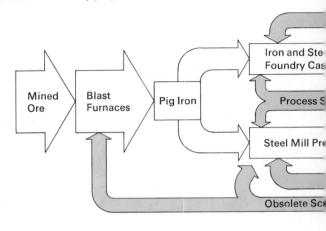

Market Classification	%	Aver
Shipbuilding & Marine Equipment	99·9	
Rail Transportation Equipment	86	
Contractor's Products	87	
Foundry	99·9	
Ordnance & other Military Equipment	36	
Electrical machinery & Equipment	75	
Mining, Quarrying & Lumbering	91	
Machinery & Industrial Tools	94	
Agricultural Equipment	99	
Containers	13	
Automotive	99·9	
Other domestic & Commercial Equipment	57	
Oil & Gas Drilling Equipment	99·9	
Appliances, Utensils & Cutlery	76	
Aircraft	99·9	

Recoverable Scrap Percent

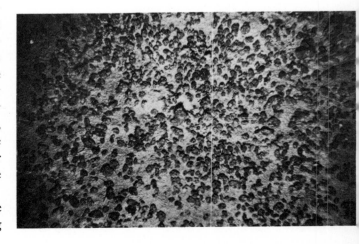

below One cubic mile of sea water, weighing about 4,700 million tons, contains 35,000 parts per million of dissolved solids. The main elements are chlorine and sodium (the constituents of salt), and only a few are today extracted in large amounts. But the potential is great, as the table (stating the number of tons of each element per cubic mile of sea water) shows.

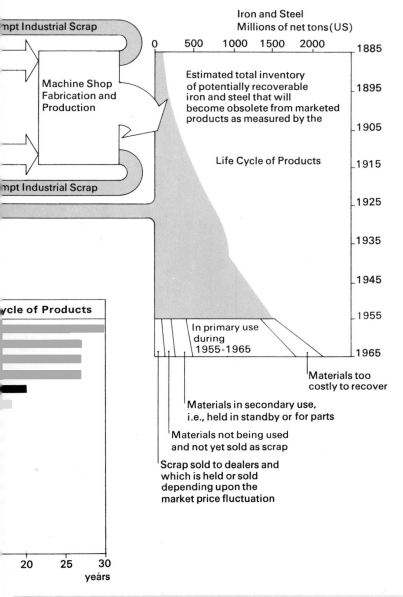

Iron and Steel
Millions of net tons (US)

Estimated total inventory of potentially recoverable iron and steel that will become obsolete from marketed products as measured by the

Life Cycle of Products

In primary use during 1955-1965

Materials too costly to recover

Materials in secondary use, i.e., held in standby or for parts

Materials not being used and not yet sold as scrap

Scrap sold to dealers and which is held or sold depending upon the market price fluctuation

Machine Shop Fabrication and Production

mpt Industrial Scrap

cycle of Products

20   25   30
years

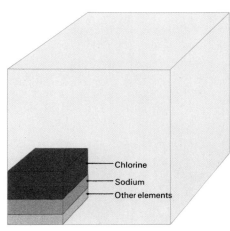

Chlorine →
Sodium →
Other elements →

Chlorine	89,500,000		Nickel	9
Sodium	49,500,000		Vanadium	9
Magnesium	6,400,000		Manganese	9
Sulphur	4,200,000		Titanium	5
Calcium	1,900,000		Antimony	2
Potassium	1,800,000		Cobalt	2
Bromine	306,000		Cesium	2
Carbon	132,000		Cerium	2
Strontium	38,000		Yttrium	1
Boron	23,000		Silver	1
Silicon	14,000		Lanthanum	1
Fluorine	6,100		Krypton	1
Argon	2,800		Neon	0·5
Nitrogen	2,400		Cadmium	0·5
Lithium	800		Tungsten	0·5
Rubidium	570		Xenon	0·5
Phosphorus	330		Germanium	0·3
Iodine	280		Chromium	0·2
Barium	140		Thorium	0·2
Indium	94		Scandium	0·2
Zinc	47		Lead	0·1
Iron	47		Mercury	0·1
Aluminium	47		Gallium	0·1
Molybdenum	47		Bismuth	0·1
Selenium	19		Niobium	0·05
Tin	14		Thallium	0·05
Copper	14		Helium	0·03
Arsenic	14		Gold	0·02
Uranium	14			

far left Manganese nodules from the sea bed : These potato-sized lumps of rich ore are coagulated out of sea water, and are now being harvested with machines like giant vacuum cleaners. left Salt has been harvested from the sea for thousands of years, using the Sun's heat to evaporate the water. Bromine is extracted from the liquor that remains.

114

Technological developments in the future will make it much easier for people to go into the underwater environment. **right** Marine archaeology, still a young science, can expect to expand even further.
below With increasing leisure, underwater exploration may become as popular as boating and sailing are today. **opposite** The biggest impact of technology is, however, likely to be in the deep waters; working in these surroundings will be an adventure as well as a job. This artist's vision shows pipe layers of the future with their underwater tractor.

wind power, sunlight conversion, tidal power and heat layer differentials in the sea. It would be a very interesting move for man to be forced to return to the freely available and 'natural' energy sources. Such a switch would have to be accompanied by some sobering reassessments of physical needs and social structure.

The ocean ethic

One of the planet's earliest successful cycles that settled into an equilibrium, given the intervals of the glacial rhythms, was the hydrocycle. In his inevitable move into the oceans, it is very much to be hoped that mankind will learn this wisdom of the water, this cyclicity, as the basis of his own industrial behaviour. If mankind continues to attempt to deny the first law of life – cyclicity – and does not put checks on his discarded artifacts, whether by-products of their manufacture or solid, liquid or gaseous refuse, he will upset an equilibrium that has taken millions of years to settle – possibly irreversibly. This is the responsibility of his new knowledge. It therefore becomes obvious that future oceanographers, or oceanmen, will have to be generalists steeped in practical common-sense and highly sensitive of their vital role as guardians of not only man's but the whole planet's last resource.

It is for this reason that the oceanmen will have to be educated on a very broad front, not only embracing geology, meteorology, limnology, seismology, biology, geophysics, chemistry, instrumentation, underwater acoustics, engineering, and physical oceanography, but also the overview sciences such as general systems theory, cybernetics, philosophy, and personal and collective survival systems. They will have to acquire a balance of cultivated knowledge tested in action and rooted in a value system which starts with the whole, emphasizing not resources to man's advantage but resourcefulness to benefit the whole system of which man is an essential part.

Whether watching sonar screens to monitor fish shoaling, counting the world's protected whales, keeping a check on pollution levels or tracing their sources, or even surveying by helicopter – will need a range of skills. Other careers, such as managing a coastal bay fish-farm, or even colonizing floating oceanographic stations or migrating aquacultural filter farms, will become increasingly available in the future. Others will prefer the underwater world of excitement, danger and exploration. In whatever capacity, the ocean environment is the world where man may find new opportunities and rediscover old truths about his relationship with the world he inhabits.

Chapter 6

Oceanic Bill of Rights

MAN has been described as the eyes of the planet. It is then a direct, logical progression to consider him as the planet's nervous system – its voluntary, conscious nervous system. It is hardly likely that the fate of so subtle a product as the eyes is a matter of indifference to the system that evolved them. Yet it is certain that the human component can choose to destroy the immense experiment of life – either by a quick radioactive route or a slow consumption-pollution route.

Of the first, little can be said except to repeat the fact that nobody could win a nuclear holocaust; biological damage would be so extensive as to destroy the capacity of the rest of the planet to support anything that remained. As for the second, it is in just this realm of pollution that the oceans become the decisive factor. The reason has been demonstrated; it is because the oceans are not only life's sewage system and its water supply but also its thermostat and its oxygen producer.

What is at the root of the problem? At a very basic level, nature and life are governed by the law of balance. A living system is one of very fine balance. What is taken in to be utilized must be balanced by an output. The organism accomplishes its self-directed aims with the 'borrowed' energy. If the law of balance is applied to the oceans, we see very clearly that whatever is taken out or put in is distributed in a number of ways, which have already been outlined. Always the forces tend towards a distributive equilibrium.

Landmarks in the modern history of man have been embodied in the constitutions of new states and major human organizations like the United Nations. These have been formed with the rights of men and affairs between nations uppermost. An organism obviously cannot develop without some internal coordination. Now is the time, before the negative forces accelerate beyond the point of no return, for us to consider the rights of the ocean. What are the principles to be maintained for health and survival of the whole system? If a constitution for the oceans – an oceanic Bill of Rights – were to be formulated, what would be its basis? In human terms, what is necessarily put back correctly we call duties, what is taken out freely we call rights. So from the point of view of the oceans we could ask: What are the rights that the seas can demand of men – if men are going to continue to be supported? A draft Bill of Rights embodying these principles is shown opposite.

Just as it took the crisis of World War II to galvanize the separate peoples of the world into forming the United Nations and collectively signing the Universal Declaration of Human Rights, it will probably be an ecological crisis that will secure an attitude that observes those environmental rights which support the existence of life at source. It is plain that man has to regain a psychological harmony with the other functional elements of the planet – a 'work with' mentality instead of an 'exploitation' mentality. It will be essential to replace 'the survival of the fittest' as a human or species concept with a total-systems concept of 'the survival of the fittest to cooperate'.

Cooperation or exploitation?

It is essential to constantly remind ourselves of the value of harvesting the seas – providing we take to heart the lessons of the misuses of our soils, and legislate against the same problems occurring in the far more delicate oceanic environment. However, once one talks of legislation it is also essential to be realistic and well informed as to what the great powers are currently doing in the seas.

Both the Soviet Union and the United States accepted the resolution adopted by the General Assembly of the United Nations on 14 January 1961 which contained the statement that the Assembly 'reapproved that exploration and exploitation of the reserves of the sea-bed and the ocean floor, and the subsoil thereof, should be carried out for the benefit of mankind as a whole, taking into special consideration the interests and needs of the developing nations', and 'recalled that international cooperation in this field is of paramount importance'. Yet on 5 January 1968, the *Philadelphia Inquirer* spoke of a U.S. Navy-supported industrial project: 'Working under Navy contract, researchers here are studying ways to house up to 1,000 men on the deep ocean floor to establish advanced undersea warfare systems.' Typifying the mentality of a divided world, Dr. Athelston F. Spilhaus, Dean of Minnesota's Institute of Technology, wrote in 1964: 'Man is going to colonize the oceans, and it might as well be *our* men.'

A glance at these approximate figures for U.S. government expenditure on the oceans tells its own story:

	1964	1970
Defence	$454 million	$1,000 million
Civil	$127 „	$ 300 „
Waste treatment	$335 „	$ 305 „
Totals:	$916 million	$1,605 million

The defence expenditure is nearly twice the civil and waste treatment budgets put together. Some members of the

right The 1971 Malta conference on the peaceful uses of the sea brought together such distinguished figures as Malta's Arvid Pardo (*third from left*) and Norwegian anthropologist Thor Heyerdahl (*far left*). The meeting signified some hope for a more responsible attitude to the seas. **below** A draft for an Oceanic Bill of Rights.

Oceanic Bill of Rights

WHEREAS the peoples of the world recognize:

that the oceans of planet Earth have played and continue to play a unique and fundamental role in the creation, nurturing and maintenance of life;

that the oceans have established a state of balance and equilibrium, in both physical and biological terms, of benefit to all components of the planetary system;

that mankind, through technological developments, has acquired the ability to make or break the delicate oceanic balance;

that the oceans can yield great material and living resources of value to mankind and other members of the living world, provided that the universal laws of balance are observed;

and that man is a unique component of the planetary system, possessing consciousness, foresight and wisdom;

THEREFORE let it be enacted as follows:

1 The oceans of the world have the inalienable right to be free to support life.

2 The oceans have the right to be unadulterated by alien substances designed by man which impair their natural functions.

3 The oceans have the right to demand of all who are supported by them the payment of correct and immediate dues for what they extract in terms of maintaining correct oceanic balance.

4 The oceans have the right to demand that all men who assume guardianship for areas of the ocean shall observe their fellow men's inalienable rights of freedom, justice and peace, and shall refrain from using such areas as a theatre of conflict under any context whatsoever.

5 The oceans have the right to expect that all mature life deliberately taken from their guardianship shall be returned in equal volume in young life either in the form of human-nurtered spawning or human-monitored sea cultivation.

6 The oceans have the right to demand that whosoever takes the liquids of their bulk in any form for whatever function shall return liquids with a marine-digestible content that benefits sea life.

7 The oceans have the right to be understood. As a basis for all the above rights, the oceans have the right to demand that man shall use all his powers of wisdom to understand, and to transmit his understanding to all his fellow men, all the details of the vital and simultaneous functions the oceans perform in the phenomenon of life on Earth.

Pentagon are blunt about their commitment. Admiral 'Muddy' Waters wrote in the defence industry bulletin: 'Oceanography is not to be confused with anti-submarine warfare, nor the Polaris system, nor amphibious or mine-warfare operations. Oceanography is a necessary support element in all of the warfare areas.' As if this were not pointed enough, Robert H. Baldwin, Under Secretary of the Navy, went further: 'We are involved in deep ocean engineering because it contributes to our assigned missions; we are not in the business of exploring the ocean's abundant mineral or living resources.'

The deployment of nuclear weapons in the oceans is an established fact. At least four nations – the Soviet Union, the United States, Britain and France – have or are developing fleets of nuclear-powered submarines armed with nuclear ballistic missiles. Recent estimates suggest that the number of such submarines is rapidly overtaking the 100 mark. And for every submarine there are submarine-hunters. In the words of Malta's eloquent United Nations representative, Arvid Pardo, who has done so much to advance peaceful uses of the oceans, 'An informed guess is that the United States Navy is currently spending about £400 million for submarine tracking and detection devices installed on the ocean floor.'

Not all American politicians are taken in by government claims as to the value of U.S. Navy monopoly in oceano-graphy, as Senator Clairborne Pell of Rhode Island, a pointed spokesman, has made clear: 'Our national dependence on the Navy for the major sponsorship of ocean development has left gaps in the national program . . . We trail the Soviet Union in fisheries, we are far behind the Japanese in aquaculture, and we have left to the French most of the development of systems to obtain sea power.' Still further reasoned thinking is beginning to see an element of nightmarish fantasy behind the whole arms-race policy. Dr. J. Wiesner put this very well in the journal *Industrial Research* in July 1969. 'We are running,' he said, 'an arms race with ourselves, and very frequently the threats that we can see five years ahead, for example, are the threats that we ourselves pose.'

Patience, well-disseminated information and a cool head are needed to foresee the time when the admirals, generals and politicians will accept the findings of their war-gaming computers and face up to the fact that there is no 'winning' to be done in a nuclear war. At the same time a glimmer of hope can be seen in the immense competitive survival pressures that are accelerating oceanographic research – ridiculously duplicated – through an artificially-generated fear strategy. If our previous rates of acquiring academic knowledge were anything to go by, we probably would not have understood the sea properly before our pollutants had done irreparable damage.

As it is, it may ironically be the very same secret information, which the great powers pledged in their U.N. resolution to disseminate, that will precipitate a knowledge of the ecological crisis which in turn will force a new attitude. Men may differ, even destroy each other, but if the oceans are no longer able to supply the air through phytoplankton-overkill, an even more foolish mankind will die while allegedly defending himself against himself. Apart from the obvious dangers of a nuclear accident with so many warheads being trailed around the seas, the radioactive leakage – however small – from these nuclear-powered submarines is perpetually increasing the oceans' loading. The oceanic monitoring work these submarines are also obviously engaged in may well spell out their own extinction as a source of fatal oceanic pollution.

New laws for the seas

Meanwhile, every effort must be made to face up to the inevitable next step: that of drawing up a code of practice to ensure at least a basis for tolerable behaviour on and in the seas. Dedicated men such as Arvid Pardo and Professor W. T. Burke, of Columbus, Ohio, have been working on models for a legal structure for the world oceans, as have people such as Laurence Reed, whose excellent booklet *Ocean Space: Europe's New Frontier* has approached the subject on a smaller scale.

The issues are large and fundamental. So far, the majority of the planet does not come under the heading of being 'owned' by men. In fact the concept of men owning areas of the planet is by no means universally accepted within mankind itself, and is even cited by some as being the most consistent source of conflict between men. So the first question is: Should some form of international regime representing mankind's interests, as defined by the U.N., own the ocean bed beyond the sovereign limits of nations? Or would it be a timely opportunity to establish the more sound ecological position of a regime of guardians whose clients were mankind and whose ethics were the oceanic Bill of Rights? This would re-establish a fundamental concept of many traditional nomadic peoples – man as guardian of the part of the Earth he occupies, not as owner of it.

Next, what role and form should this intergovernmental guardianship take? It would be foolish and naive to over-

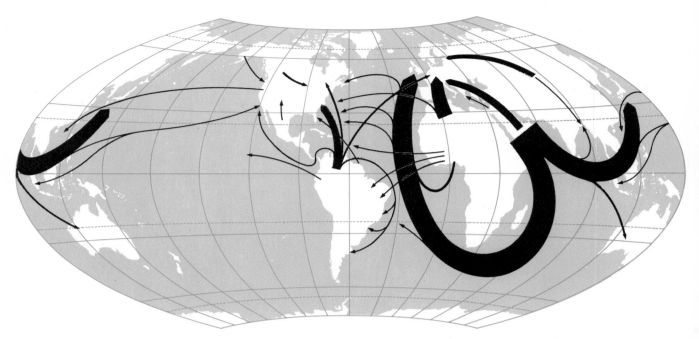

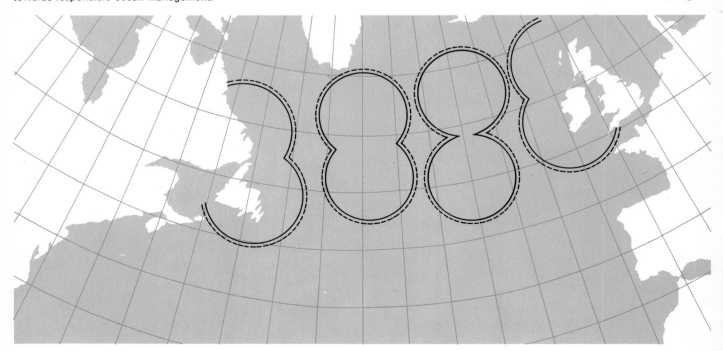

above The international movements of oil and gas – highly influential international commodities on which so much of industrial society relies – illustrate the problems involved when national commercial interests become involved with attempts to introduce international controls. **opposite top** The carve-up of the North Sea into blocks for oil and gas exploration epitomizes irresponsible marine exploitation. It represents the extension of man's appetite for ownership from the land to the previously un-'owned' seas. **opposite bottom** A comparison of American and Russian naval strengths emphasizes the great powers' overriding oceanic interest – military control. **below** This plan for a series of floating ocean monitoring platforms shows a movement towards responsible ocean management.

simplify the answers to this question but it is generally agreed that certain urgent measures must be taken immediately while the correct formula is worked out. First, there must be a freeze or moratorium on present claims to ocean bed areas by sovereign states, limited to 12 miles off-shore of any state. Second, all activities, claims, installations, etc, so far in existence must be registered, the act of registration in no way ratifying or legalizing the activity. Third, there must be a reclassification of the continental shelves based on mileage, depth or a geological concept, so as to prevent unlimited extensions of exclusive claims.

It would be a historically tragic error not to grasp the opportunity of establishing at a stroke the universal rights

of all men to the ocean's benefits, in the same way that the ocean universally benefits all men by its donation of water supply. The major task of the agency set up to exercise guardianship over the oceans would be the establishment of principles to govern the four major issues of practical usage: scientific research into the oceans, military use of the sea bed, acquisition of marine fishery resources, and acquisition of mineral resources.

The most difficult questions arise out of the universal claim of all men to benefit from the oceanic harvests. Should the agency make a once-for-all distribution of titles to parts of the ocean bed, collecting 'rent' from those drawing the leases? Is there any other distribution method besides population proportion? What charges on leases should be made both initially and for running costs? In what way should the funds accruing from this international rent-collecting be used? To finance the agency's research and exploration, surveillance and data distribution? To finance the United Nations? To help poorer countries or to be distributed to all countries in proportion to population? Should the agency take the responsibility of enforcing the recording and publishing of information on resources extracted from the oceans as a compulsory condition of the lease? Should it also establish the freedom of continental shelf research so long as national programmes are declared? Should it be responsible for publishing information on all installations, structures and apparatus erected or laid on the sea bed outside international waters?

Finally, the question arises of maintaining public order on the oceans. What role should the agency play? It would seem wisest for it to give factual information directly to the United Nations, making it essential that a U.N. oceanic police force be instituted to enforce agreements. Then the agency would be left free to play its role as peaceful developer of oceanic resources and as oceanic ombudsman.

Man's industrial ingenuity has resulted in many people reaching a level of comfort far beyond the satisfaction of basic human needs. But it has left the world in a unique state of economic imbalance. For the comfortable, the recent discovery of global ecological rules has come as quite a shock. The oceans are as sensitive an organ of the ecological imbalance as any. At this moment in time there is a unique opportunity for mankind to institute, on the largest scale ever, the principle of human guardianship. He can establish conscious, interdependent cooperation with the fellow forces of the planet in a concerted attempt to rediscover the right relationship for humanity to continue its role within the organism Earth.

USA	USSR
14 attack carriers	
10 helicopter and support carriers	2 helicopter carriers
9 cruisers	25 cruisers
214 destroyers, frigates and destroyer escorts	190 destroyers, frigates and destroyer escorts
95 nuclear-powered submarines	90 nuclear-powered submarines
42 other submarines	260 other submarines
68 landing craft	100 landing craft
2 torpedo and missile boats	560 torpedo and missile boats

Oceanaut Farooq Hussain with dolphin
friend: Only through understanding can man
get back into a harmonic relationship with
the oceans and their inhabitants.

Index

Picture credits

The Editors gratefully acknowledge the courtesy of the following artists, photographers, publishers, institutions, agencies and corporations for the illustrations in this volume.

Front Flap
David Nockels
Title Page
B. Griffith
Page
8 Eugene Fleury
8–9 Keith Critchlow
9 Tom McArthur
10–11 David Nockels
11 Sólarfilma, Reykavik, Iceland
12–13 Eugene Fleury
14 Keith Critchlow
15 David Nockels
16 Tom McArthur
   Eugene Fleury
17 Tom McArthur
18 Keith Critchlow
19 Tom McArthur
   Eugene Fleury
   Eugene Fleury
20–21 Vana Haggerty
   Life © Time Inc.
   Photographer: Arthur Rickerby
   Professor J. G. Ramsay
22–23 Aerofilms
24 Keith Critchlow
25 Tom McArthur
   Eugene Fleury
   Eugene Fleury
26 Eugene Fleury
   By courtesy of NASA
27 Eugene Fleury
28–29 David Nockels & Eugene Fleury
30–31 Eugene Fleury
32–33 David Nockels & Eugene Fleury
34 Courtesy of the Shell International Petroleum Co. Ltd
   By courtesy of NASA
   Photo: courtesy of the Marine Physical Laboratory, Scripps Institution of Oceanography, University of California
35 Les Requins Associé, Paris
   Scipps Institution of Oceanography
   Naval Photographic Center, Washington D.C.
   Courtesy of Vickers Limited
36 Eugene Fleury
36–37 Eugene Fleury & David Nockels
38 Keith Critchlow
   Eugene Fleury
38–39 David Nockels
40 David Nockels
41 Tom McArthur
42 Tom McArthur
42–43 David Nockels
44–45 David Nockels
   Eugene Fleury
46–47 David Nockels
49 Eugene Fleury
   David Nockels
50–51 David Nockels
52–53 Jack Fields
   David Nockels
54–55 David Nockels
56–57 David Nockels
58 Tom McArthur
59 David Nockels
61 David Nockels
   Eugene Fleury
62–63 Eugene Fleury
   David Nockels
64 Eugene Fleury
65 Eugene Fleury
66–67 Bruce Coleman/Photo: Tom Weir
67 Eugene Fleury
68–69 Eugene Fleury
   By courtesy of the Trustees of the National Gallery, London
70 Keith Critchlow
   Tom McArthur
72 David Nockels
73 David Nockels
74 Keith Critchlow
   Tom McArthur
   Keith Critchlow
76–77 By courtesy of NASA

78 Tom McArthur
79 Tom McArthur
80 Heather Angel
81 Eugene Fleury
   Tom McArthur
82 Francesco Leoni
83 Eugene Fleury
84 Eugene Fleury
   Associated Press
86 Press Association
86–87 Eugene Fleury & Neil Hyde
88 By courtesy of the French Government Tourist Office, London
   Keith Critchlow
90–91 Pierre Pittet
   Naval Photographic Center, Washington D.C.
   By courtesy of the U.S. Information Service, London
92 The Detroit News
93 Eugene Fleury
94–95 S. G. Brown
96–97 Keith Critchlow
98–99 Keir C. Campbell
   Keith Critchlow
100–01 Keith Critchlow
102 Keith Critchlow
   Eugene Fleury
103 Keith Critchlow
104–15 Camera Press
   Picturepoint
106 By courtesy of Pilkington Bros. Limited
   Keith Critchlow
107 Eugene Fleury
108–9 Les Requins Associé
   By courtesy of the U.S. Information Service, London
110–11 Seaphot: Photo Peter Scoones
   Eugene Fleury
112 Tom McArthur
113 National Institute of Oceanography, Wormley
   J. Allan Cash
114 Seaphot: Photo Peter Scoones
   Seaphot: Photo Peter Scoones
115 David Nockels
116 Léonie Schilling Nagel
117 Robin Dodd
118–119 Dr C. Swithinbank
120 Eugene Fleury
121 Eugene Fleury
122 Jim Condren